Safer
Complex
Industrial
Environments

A Human Factors
Approach

Safer Complex Industrial Environments

A Human Factors Approach

Edited by Erik Hollnagel

CRC Press
Taylor & Francis Group
Boca Raton London New York

CRC Press is an imprint of the
Taylor & Francis Group, an **informa** business

CRC Press
Taylor & Francis Group
6000 Broken Sound Parkway NW, Suite 300
Boca Raton, FL 33487-2742

First issued in paperback 2017

© 2010 by Taylor and Francis Group, LLC
CRC Press is an imprint of Taylor & Francis Group, an Informa business

No claim to original U.S. Government works

ISBN-13: 978-1-4200-9248-6 (hbk)
ISBN-13: 978-1-138-11680-1 (pbk)

Library of Congress Cataloging-in-Publication Data

Safer complex industrial environments : a human factors approach / editor, Erik
   Hollnagel.
       p. cm.
     "A CRC title."
     Includes bibliographical references and index.
     ISBN 978-1-4200-9248-6 (hardcover : alk. paper)
       1. Human engineering. 2. Industrial safety. 3. Industrial hygiene. I. Hollnagel, Erik,
   1941- II. Title.

TA166.S245 2010
363.11--dc22   2009027799

Visit the Taylor & Francis Web site at
http://www.taylorandfrancis.com

and the CRC Press Web site at
http://www.crcpress.com

Contents

Preface

The International Symposium on Human Factors 2007 was held November 22 and 23, 2007, in Mihama-cho, Fukui, Japan. It was a commemorative symposium marking the fifteenth anniversary of the establishment of the Institute of Nuclear Safety System, Inc., and also the tenth anniversary of its relocation to Mihama. This publication is based on a selection of papers presented at the symposium.

The Institute of Nuclear Safety System (INSS) was established in 1992 by the Kansai Electric Power Company, Inc. The purpose was to conduct comprehensive studies in order to improve the level of safety and reliability of nuclear power plants. INSS is comprised of two institutes: the Institute of Social Research and the Institute of Nuclear Technology. The results of their research activities have been published worldwide.

INSS established the Research Center for Human Factors in 2007 as the core of the Institute of Social Research. Its mission is to conduct research activities beyond the conventional framework to understand human factors more broadly and thereby to contribute to preventing human errors more effectively.

The objectives of the symposium were to share knowledge and experiences among the world's leading researchers in various industrial fields concerning human factors, including safety climates of organizations, and to discuss the direction of future research activities.

It has been found by INSS and others that accidents and problems, in the nuclear and other industries, are caused not only by human errors but also by issues of safety awareness and organizational problems. Future research must therefore embrace a new framework of ideas in order to fully comprehend human factors.

A total of sixteen cross-industrial oral presentations addressed the current status and future trend of "human factors" from two viewpoints: considering safety and considering organizations. Approximately two hundred researchers and front-line staff engaged in work on safety attended the symposium and participated in active discussions.

The symposium clarified the current state of understanding and the targets of future research on human factors. We hope that the results of the symposium will be beneficial to all and help to improve the safety of nuclear power and guide research on human factors.

Yosaku Fuji
President, Institute of Nuclear Safety System, Inc.
Senior Advisor, Kansai Electric Power Company, Inc.

Contributors

Yosaku Fuji joined the Kansai Electric Power Company in 1960 after graduating from the School of Electrical Engineering, Kyoto University. He became the company's president in 2001. In 2006, he became the president of the Institute of Nuclear Safety System, Japan, an institution that conducts research for the improvement of safety and reliability of nuclear power generation.

Hirokazu Fukui is a senior researcher of the Research Center for Human Factors at the Institute for Social Research, the Institute of Nuclear Safety System in Japan. After joining the Kansai Electric Power Company in 1975, he worked mainly in a maintenance section for machinery in a nuclear power plant until leaving for the INSS in 1997. His research interests are safety culture and organizational learning in nuclear power plants.

Yutaka Furuhama has been a researcher with the Human Factors Group at Tokyo Electric Power Company (TEPCO) since 1998. He received a PhD in engineering from the University of Tokyo in 1996, joined TEPCO in the same year, and worked at Kashiwazaki-Kariwa NPP for two years. His recent activity includes improvement of the RCA method, analysis and promotion of organizational safety, and development of instructional materials for human factors.

Takaya Hata holds a bachelor's degree in high energy nuclear physics from Tokyo University of Science. He joined Japan Systems Corporation in 1983. After working mainly in the simulator model development section, PSA section, and emergency response support section in the nuclear power plant, he became the senior researcher of the Human Factors Evaluation Group, Safety Standard Division, at the Incorporated Administrative Agency, Japan Nuclear Energy Safety Organization (JNES).

Yuko Hirotsu holds a bachelor's degree in engineering from Keio University (Japan). He joined the Central Research Institute of Electric Power Industry (CRIEPI) in 1992. His main research area is analyzing events due to human errors at electric power industries.

Erik Hollnagel is a professor and the industrial safety chair at MINES Paris-Tech (France), visiting professor at the Norwegian University of Science and Technology (NTNU) in Trondheim (Norway), and professor emeritus at the University of Linköping (Sweden). Since 1971 he has worked at universities, research centers, and industries in several countries and with problems from several domains, including nuclear power generation, aerospace and aviation, software engineering, healthcare, and land-based traffic. His professional interests include industrial safety, resilience engineering, accident investigation, cognitive systems engineering, and cognitive ergonomics.

Shang H. Hsu received a PhD in psychology from the University of Georgia in the United States. He is a professor of industrial engineering and management at National Chiao-Tung University, Hsinchu, Taiwan. His research areas include human–machine interface, safety management of complex systems, and cognitive engineering.

Yoichi Ishii holds a master's degree in electrical engineering from Chuo University. He joined the TEPCO (Tokyo Electric Power Company) in 1982. After working mainly in the operations section and engineering section in nuclear power plants, he became the senior officer of the Human Factors Evaluation Group of JNES (Japan Nuclear Energy Safety Organization).

Masaharu Kitamura was born in Morioka, Japan, and obtained his bachelor's, master's, and doctoral degrees from the School of Engineering, Tohoku University, Sendai, Japan. In 1992 he became professor at the department of nuclear engineering at Tohoku University, after serving as a postdoctoral research associate there and at the University of Tennessee, Knoxville, in the United States. He is now professor emeritus and also director of the organization management project at the New Industry Creation Hatchery Center (NICHe). His main areas of research are instrumentation, control, human–machine interface technologies, human factors, public communication, and ethics in engineering.

Pierre Le Bot has extensive experience in nuclear power plant safety engineering in general and in accident prevention and mitigation in particular. He has been an expert researcher in human reliability analysis for EDF's R&D since 1993. First, he contributed to the human data collection from observations on simulators for HRA method development and application. He then led the development of MERMOS (*méthode d'evaluation de la réalisations des missions opérateur pour la sûreté*), currently being implemented at EDF. A former graduate in sociology from the Institute of Political Sciences (IEP, Institut d'Etudes Politiques) in Paris, France, he is currently focusing his research on the impact of organizations on human reliability and has recently developed the Model of Safe Regulation, which proposes a theoretical modeling of resilience for engineering of risky systems.

Chun-Chia Lee is a PhD candidate of the Department of Industrial Engineering and Management, National Chiao-Tung University, Hsinchu, Taiwan.

Maomi Makino holds a master's degree in electrical engineering from Kyoto University. He is a director of the Human Factors Evaluation Group at the Incorporated Administrative Agency, Japan Nuclear Energy Safety Organization (JNES). Before that, he was a general manager of the Institute of Human Factors in the Nuclear Power Engineering Corporation (NUPEC). He has been engaged in research and regulatory support on human and organizational factors, safety culture, and human–system interface.

Atsushi Naoi is director of the Institute of Social Research, Institute of Nuclear Safety System. He is an executive member of the Science Council of Japan. His main fields are the sociology of social inequality and work, and the social impact of information technology. At present, he has strong interest in safety culture and the theory of high-reliability organization.

Makoto Ogasawara joined Mitsubishi Atomic Power Industries in 1972, working mainly in the electrical and control engineering section of the water reactor division. Now he is the senior officer of the Human Factors Evaluation Group of the Safety Standard Division at the Incorporated Administrative Agency, Japan Nuclear Energy Safety Organization (JNES) and has been engaged in the study of human and organizational factors.

Hiroshi Sakuda holds a master's degree in electrical engineering from Doshisha University. He joined the Kansai Electric Power Company in 1980. After working mainly in the quality assurance section in nuclear power plants, he became the project manager of the Human Factors Project in 2000 and the director of the Research Center for Human Factors in 2007 at the Institute of Nuclear Safety System.

Hiroyuki Shibaike graduated from the Osaka Prefecture University in 1988 and entered the Kansai Electric Power Company. At the company, he has been engaged in tasks related to the maintenance of nuclear power plant facilities and the treatment and disposal of radioactive waste, and is presently involved in human performance improvement activities at nuclear power plants.

Masayoshi Shigemori is the senior researcher in the Safety Psychology Laboratory, Human Science Division, Railway Technical Research Institute. He earned his bachelor's degree from Rikkyo University in 1991, and a master's from Gakushuin University in 1994 in cognitive psychology. Research interests include action slips and interference of information processing within the brain.

Toshio Sugiman is a professor at the Graduate School of Human and Environmental Studies, Kyoto University, Japan. He has carried out action research in various fields such as communities, organizations, disaster prevention, environmental conservation, technological development, international conflicts, etc., including nuclear power industries. He was president of the Japanese Group Dynamics Association from 1994 to 1998 and was awarded a fellow of the International Association of Applied Psychology in 2006.

Kenichi Takano received his PhD in nuclear engineering from Nagoya University. He is now a professor at the School of System Design and Management, Keio University, Tokyo, Japan.

Muh-Cherng Wu received his PhD in industrial engineering from Purdue University in United States. He is a professor of industrial engineering and management at the National Chiao-Tung University, Hsinchu, Taiwan. His research areas include production management and CAD/CAM.

Michio Yoshida is a professor at Kumamoto University and president of the Japan Institute for Group Dynamics. He has been engaged in developing leadership and human relations training. He is also interested in preventing accidents and establishing a safety culture in organizations. He received his PhD from Hiroshima University.

1 The Background and History of the Institute of Nuclear Safety System (INSS)

Atsushi Naoi

CONTENTS

INTRODUCTION

The Kansai Electric Power Company, Inc. (KEPCO) operates eleven nuclear power reactors in Fukui prefecture, Japan, which generate approximately 40% of Japan's total power supply. In accordance with the laws and regulations of Japan, the number of forced outages at nuclear power plants must be reported to the authorities. For KEPCO, the reported number was 0.14 outages per reactor year in the ten-year period from 1998 to 2007, which must be considered a relatively good operating record.

1

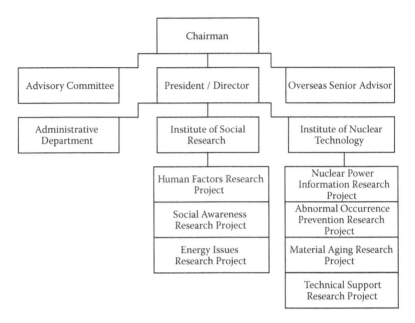

FIGURE 1.1   Original organizational structure of INSS as of March, 1992.

The Institute of Nuclear Safety System, Inc. (INSS) was established by KEPCO in March 1992. The institute was created in the aftermath of a steam generator tube rupture accident at KEPCO's Mihama Power Plant Unit 2 that had occurred in February 1991 (see the account given in Chapter 2). The accident, which was rated as Level 2 on the International Nuclear Event Scale, had a great impact on society at that time because it was the first time the emergency core cooling system had been activated in a Japanese nuclear power plant. The accident was clearly associated with human factors, suggesting the need for an approach that considered both the technical and social safety of nuclear power generation. Two institutes, the Institute of Nuclear Technology and the Institute of Social Research, were therefore created as part of the INSS, with the aim of doing extensive research from the perspectives of both nuclear engineering and human or social sciences, and thereby helping to achieve greater harmony between society and the environment. The organizational structure of INSS is shown in Figure 1.1.

The INSS is built on three fundamental principles that serve to guarantee its independence. The first principle is to carry out extensive research from both technological and social aspects. The second principle is to do objective research from an independent or third-party standpoint and to raise constructive suggestions for the development of nuclear power generation. The third principle is to conduct open research activities and make the outcome of research available to as many people as possible. Based on these fundamental principles, the INSS has energetically pushed ahead with various research projects, the results of which have been highly appreciated both at home and abroad. Some of these results can be found in Misumi, Wilpert, & Miller (1998), Wilpert & Itoigawa (2001), and Itoigawa, Fahlbruch, & Wilpert (2004).

These efforts notwithstanding, another serious accident happened when a secondary system pipe ruptured at KEPCO's Mihama Power Plant Unit 3 on August 9, 2004.

In this accident five people died and six others sustained serious injuries. Needless to say, it came as a great shock to the INSS, which was vested with the mission of assuring safety.

AN OVERVIEW OF THE MIHAMA-3 ACCIDENT

The rupture of a secondary system pipe occurred on August 9, 2004, while the Mihama Unit 3 was in continuous operation at the rated thermal output. Steam was emitted from the secondary coolant circuit (the location of the rupture is indicated in Figure 1.2), but there was no effect from radioactivity in the surrounding environment. The main pipes of the primary coolant circuit are made of stainless steel and are about one hundred times more resistant to corrosion than the carbon steel pipes used in the secondary coolant circuit.

The chronology of the accident was as follows:

15:22   The accident occurred and a "fire alarm activated" alarm was issued.
15:25   The operators found that the third floor of the turbine building was filled with steam.
15:26   The operators concluded that the accident might be caused by a leakage of steam or very hot water and immediately started to reduce the output.
15:28   The reactor was automatically shut down, followed by the automatic shutdown of the turbine.

THE LOCATION OF DAMAGE

Workers from a contractor were carrying out their tasks in the turbine building, preparing for the twenty-first outage inspection. The inspection was to take place on August 14, so the accident happened just five days before the inspection. During the preparatory work, a condensate pipe near the ceiling of the second floor of the turbine building ruptured. Overheated steam with a temperature of approximately 140°C gushed out. Eleven of the contractor's workers present were hit by the steam and taken by ambulance to the hospital. Five of them died and six were seriously wounded.

THE CONDITIONS OF THE PIPE RUPTURE

After the accident, a large rupture or opening was found in the downstream section of the condensate pipe flow meter (the orifice). The thickness of the pipe that ruptured was originally 10 mm when the plant came online in 1976. Erosion-corrosion of the pipe wall presumably arose in the downstream section of the orifice and this gradually reduced the wall thickness of the pipe. At the time of the rupture, the pipe wall was approximately 0.4 mm thin at the section where the rupture occurred.

CAUSES OF THE PIPE RUPTURE

With regard to the inspection of the secondary system piping, KEPCO had been measuring the wall thickness of the pipe following a schedule developed in accordance with published guidelines for the management of wall thickness of the secondary

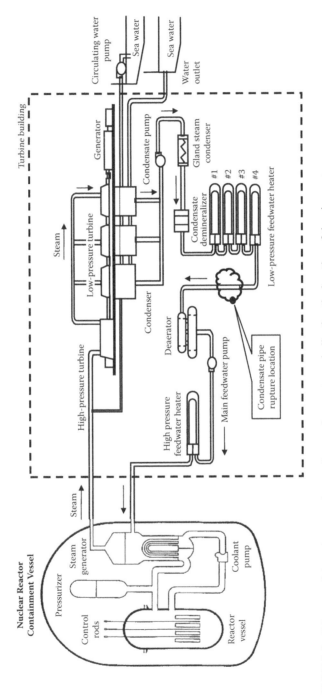

FIGURE 1.2   Diagram of the primary and secondary coolant systems, indicating the location of the pipe rupture.

system piping in nuclear installations (pressurized water reactors). The guidelines were formulated in 1990. However, the ruptured section of the pipe had initially been excluded from the list of sections to be measured. The thickness of the pipe had therefore not been measured before the accident occurred. There are several reasons why this happened.

January–June 1991   The ruptured section of the pipe was not registered in the list of pipe sections to be measured during the eleventh outage inspection. It was during this inspection that the above-mentioned guidelines were applied for the first time at the Mihama Unit 3.

1996   This omission remained undiscovered when the secondary system piping wall thickness management service was transferred from the plant manufacturer to the contractor.

1997   The omission was missed again when KEPCO commissioned the contractor to digitize inspection drawings.

April 2003   At this time the contractor realized that the section of the pipe was not mentioned in the list. The contractor, however, simply registered the section in its wall thickness management system but failed to inform KEPCO of the omission that had been discovered.

June 2003   The omission was not mentioned in an outage inspection summary report that the contractor submitted to KEPCO.

November 2003   The omission was again not mentioned in the contractor's next outage inspection plan for August 14, 2004. This section of the pipe ruptured five days before the outage.

The measurement of piping wall thickness was not part of the initial inspection routine but took place beginning with the eleventh outage inspection. At this time a list of wall thickness inspections was created for the first time, although this list did not include the pipe section that later ruptured. Altogether, the ruptured section had therefore never been inspected between the plant startup and the day it ruptured.

CORRECTIVE MEASURES

In response to this tragic accident, KEPCO implemented a number of corrective actions. The purpose of these is, of course, to prevent a recurrence of similar accidents in the future. The corrective actions are summarized in the following:

Planning and Organization of the Service
• Implementation of the secondary system piping wall thickness management service under the direct management of the company's own personnel.
• Reinforcement of on-site inspections.

- Improvements of the wall thickness management system.
- Periodical review of the pipe inspection list.
- Increase in the number of personnel engaged in the secondary system piping wall thickness management service.

Procurement Management
- Clear definition of the roles of the contractor and KEPCO.
- Clarification of the requirements, including the requirement to report omissions in the pipe inspection list.
- Reinforcement of the details to be checked and the procedure for acceptance inspection.

Information Sharing
- Designation of a leakage as a nonconformity issue under the standards.
- Definition of nonconformities other than leakages under the standards.
- Appointment of a senior information controller.
- Direct communication of crucial information to key persons.
- Improvement of the program for operating experience sharing among domestic electric utility corporations.
- Sharing of information about maintenance services between electric utilities and manufacturers.
- Sharing of operating experience and information with contractors.
- Improvement of communications with contractors.
- Improvement of on-site communications.

Education
- Dissemination of lessons learned from the recent accident and other related information throughout the company.
- Education about the importance of an operation plan.
- Improvement of education programs on the importance of procurement management.

Auditing
- A shift to an auditing procedure that focuses on the operation process designed to conduct a detailed checkup on individual tasks.

INITIATIVES TAKEN BY INSS

With the Mihama-3 accident as a background and a motivation, the INSS has turned over a new leaf and has critically assessed and discussed the ways in which research is done. From the technological perspective, the INSS has decided to push ahead with research on plant life management, whereas from the social perspective it has decided to delve into the human factor issue in the quality management system. As part of this effort, the Nuclear Power Plant Aging Research Center was created within the Institute of Nuclear Technology in July 2005 and the Human Factors Research Project was reorganized into the Research Center for Human Factors within the Institute of Social Research in July 2007 (Figure 1.3). In this way, the INSS research

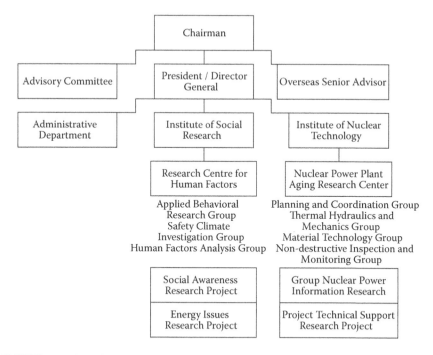

FIGURE 1.3   Organizational structure of INSS, as of July 2007.

organization has been enhanced and strengthened. The revised Institute of Nuclear Technology, the revised Institute of Social Research, and their research projects are outlined below.

THE INSTITUTE OF NUCLEAR TECHNOLOGY

Japan suffers from a scarcity of energy resources. Nuclear power covers approximately one third of its electricity requirements and thereby supports industry and national life. Because nuclear power releases only a small amount of the gases that contribute to global warming, it provides a crucial tool for protecting the environment. To be able to operate nuclear power plants in a safe and steady manner, it is essential to constantly make the necessary efforts in the field of practice. The Institute of Nuclear Technology, which consists of one center and two projects, is pressing ahead with the research projects and related activities described below, in order to further improve the safety and reliability of nuclear power plants.

Nuclear Power Plant Aging Research Center

When nuclear power plants are in operation for a long period of time, there are obviously concerns about age-related degradation of equipment and structures and about possible degradation in terms of quality assurance, maintenance, and management. The Nuclear Power Plant Aging Research Center is committed to comprehensive research into plant life management in order to prevent such potential degradation. This work is done in collaboration with the Institute of Social Research.

The Nuclear Power Plant Aging Research Center is composed of four different groups. The two first groups, the Thermal Hydraulics and Mechanics Group and the Material Technology Group, are tasked with promoting research projects for assessing and predicting the behavior of aging degradation in order to accurately determine the safety margin of equipment and structures. A third group, the Non-Destructive Inspection and Monitoring Group, is charged with promoting the development of technologies aimed to determine the extent of aging degradation (such as the size of defects). The fourth group, the Planning and Coordination Group, is engaged in planning and coordinating plant life management research projects carried out by these groups. As used here, aging degradation refers to the degradation of components that make up a nuclear power plant as they are used for a long period of time. In Japan, the first commercial light water reactor went into service in 1970, and some of the nuclear power plants have been in service for more than thirty years.

Thermal Hydraulics and Mechanics Group

Finding the causes of failures in the components and equipment of a nuclear power plant makes it possible to prevent accidents and failures at power plants. The Thermal Hydraulics and Mechanics Group carries out research projects designed to assess stress and thermal loads, one of the factors contributing to age-related degradation, and to prevent degradation-caused troubles.

The thermal-hydraulic assessment project looks at the behavior of flows that may contribute to degradation and considers options to prevent fatigue due to variations in the temperature of a fluid, cavitation damage, and vibration.

The structural assessment project assesses strength and develops analytical techniques by looking into vibrations during an earthquake and the load imposed by heat generated during the operation of a plant, using a combination of numerical analyses and experiments.

Material Technology Group

Metals used in nuclear power plants degrade due to corrosion in high-temperature, high-pressure water and due to neutron irradiation. In order to achieve the long-term safety of power plants, the material degradation mechanisms therefore need to be analyzed.

The Material Technology Group carries out mechanical tests and corrosion tests using actual and simulated materials in order to assess the durability of main metal materials for pipes and structures of nuclear power plants, such as nickel base alloys, stainless steel, and low-alloy steel. The group is also carrying out a research project designed to assess plant life as affected by aging degradation, by evaluating crack growth through material characterization and numerical analyses. This group is also playing the role of a nuclear power plant "home doctor" in surveying the causes of damage that actually occurs to equipment and piping.

Non-Destructive Inspection and Monitoring Group

To improve the safety of power plants, keeping precise track of the conditions of components and equipment in service at nuclear power plants is essential. The Non-Destructive Inspection and Monitoring Group is responsible for the development of nondestructive inspection techniques that can precisely determine the early stages of degradation of components and materials that are crucial for the safety of power plants, as well as the development of technologies for diagnosis and assessment of equipment degradation.

Under the project to develop nondestructive inspection techniques, the group is working to develop optimum automatic ultrasonic testing devices using full-scale primary coolant pipe weld-simulated specimens and to verify possible products so that they can be applied in actual systems. In addition, the group is engaged in the development of new ultrasonic testing technology capable of measuring the depth and length of defects highly accurately and technology for assessment of the mechanical properties of materials using thermal electromotive force.

Under the project to develop technologies for the diagnosis of equipment degradation, the group is working to develop technologies for the diagnosis of high-voltage motor and low-voltage electric cable insulation degradation, as well as for the assessment of the service life of polymeric materials such as rubber O-rings and diaphragms.

Planning and Coordination Group

The Planning and Coordination Group is responsible for planning, coordination, and organization of research projects conducted by the Nuclear Power Plant Aging Research Center, for gathering information about studies of plant life management, and for identifying problems to be solved. The group is also responsible for promoting collaboration with the Institute of Social Research and external research institutions.

The INSS intends to step up efforts to gather information on plant life management and collaboration with external organizations, to press ahead with research projects while identifying research topics in an unhindered manner, to achieve consistency with electric utilities' policies and external research institutions, and to find ways to contribute to actual plant life management measures at nuclear power plants.

Nuclear Power Information Research Project

A failure that occurs at a nuclear power plant may likely lead to similar events at other power plants. With this in mind, the Nuclear Power Information Research Project uses the Internet chiefly to gather information about accidents and failures at overseas nuclear power plants and then looks into and analyzes such information in order to prevent similar irregularities. For effective analyses, information gathered is categorized, processed, and put into a database. The results of the analyses are reported regularly to electric utility corporations operating domestic pressurized water reactor plants, and suggestions for improvements are submitted to them.

The project not only collects information about problematic events but also makes overseas best practices known to domestic electric power companies. It is also working to develop infrared thermography-based technologies that help prevent accidents. In addition, it is committed to research into the prevention of human errors by operators and maintenance personnel, in cooperation with the Research Center for Human Factors.

Technical Support Research Project

The Technical Support Research Project carries out research on assessment of the safety of plants and support for prevention of nuclear disasters, in order to enhance safety in the operation and maintenance of nuclear power plants.

To assess the safety of plants, the project is developing a heat transfer model in a narrow flow path at the bottom of a reactor vessel during an accident and a heat transfer model containing noncondensable gas in steam generator tubes through heat transfer hydraulics experiments. These models are used for a detailed simulation of the plant behavior based on operation and maintenance procedures. Research designed to improve operation, maintenance, and management procedures, and enrich the content of the safety management program are also undertaken.

With regard to support for nuclear disaster prevention, the project is committed to the development of assessment technology for the generation of radioactive substances, changes in the chemistry of iodine, radiation shield, and technology that predicts the progress of an event in the plant by simulating the processes leading to nuclear disaster in detail. Technology to predict radiation exposure doses to workers performing tasks in facilities during a nuclear disaster, and technology to quickly forecast the impacts of radioactive substances released from fuel on the surrounding environment, have already been developed and are employed. The behavior of a nuclear plant and the behavior of radioactive substance migration under a disaster prevention drill scenario have been analyzed and simulation data for a drill have been made available to the public.

INSTITUTE OF SOCIAL RESEARCH

Our present civilized society is supported by complicated and large-scale scientific technologies that must be safely controlled. Nuclear power generation is one of today's complicated and advanced technologies, and human beings are needed to control this technology so that it can contribute to the welfare of humankind and the development of society.

Human beings can commit various human errors either as individuals or in groups when applying the technologies and tools that are part of a modern industrial environment. Exploring methods to shed light on the human error mechanism, to predict and prevent errors, and to investigate the underlying causes of human errors is therefore crucial to controlling scientific technologies and to operating and utilizing them properly.

Scientific technologies, no matter how efficient and cost-effective they may be, mean nothing as such unless they are accepted by the general public. In order to enhance public acceptance of nuclear power generation it is crucial to keep track of the trends in public perceptions toward nuclear power. It is likewise essential to strengthen communications with the general public. In a country such as Japan, which is poor in energy resources, efforts to help people gain correct knowledge and understanding of energy and environmental issues are essential, on both a short-term and a long-term time scale.

In recognition of these conditions, the Institute of Social Research is committed to research projects that aim to achieve the safety of nuclear power generation and to establish better harmony between nuclear power and society — from the perspectives of both social and human sciences. The Institute of Social Research is composed of the Research Center for Human Factors, the Social Awareness Research

Project, and the Energy Issues Research Project. The activities of these units are outlined in the following sections (see Figure 1.3).

Research Center for Human Factors

Human errors at nuclear power plants have decreased, thanks to improvements in human–machine interfaces and the working environment, although the decrease seems to have leveled off in recent years. Methods and measures to reduce human errors ever more effectively are therefore still needed. During the early years of the 21st century the industrial world has experienced a sharp increase in problems, sometimes leading to corporate scandals that cannot be solved by the human error-reducing measures that have been used thus far. Moreover, as is evidenced by the secondary system pipe rupture accident that occurred in August 2004 at Mihama Unit 3, the need has arisen to explore the human factor issues in the quality management system. A new task has also emerged, such as analyses of underlying causes as part of a new inspection system.

To support a wider range of research topics, the Research Center for Human Factors is therefore supported by three subdivisions, which are the Applied Behavioral Research Group, the Safety Climate Investigation Group, and the Human Factor Analysis Group.

Applied Behavioral Research Group

Serving as a control tower for the Research Center for Human Factors, the Applied Behavioral Research Group is responsible for planning, coordination, and organization of human factor-related research projects, and for promoting collaboration with external institutions. The group also carries out basic research by applying psychological knowledge in order to reduce human errors.

The group has until now completed research projects on improving the guidelines for sign representation, psychological analyses of serious accidents that have occurred in other industrial sectors, and the effectiveness of a cross-check.

Safety Climate Investigation Group

This group is responsible for the improvement and standardization of the safety culture surveying methods at nuclear power plants. These methods are also used in other divisions such as thermal power plants and substations, and in other corporate organizations, as well as in consulting services for the organizations surveyed.

The Safety Climate Investigation Group has until now carried out organizational safety culture surveys (questionnaire surveys and field surveys), officers' leadership surveys, and sense of business surveys.

Human Factor Analysis Group

This group is charged with analyzing the underlying causes of human factor-related problems, conducting analysis training sessions, and carrying out practical research designed to prevent human errors.

The Human Factor Analysis Group has until now carried out research projects focusing on a human error analytical approach, on what human factors education should be like, and on the transfer of maintenance techniques and skills to younger generations.

Social Awareness Research Project

The project looks into how people perceive nuclear power generation, as a huge and complicated scientific technology whose safety must be achieved with utmost care, and how better relationships between nuclear power generation and the general public should be forged. The research projects include conducting surveys of the trends in public perceptions toward nuclear power from multiple angles, in order to solve these and other problems.

First, surveys carried out on the trends in public perceptions toward nuclear power include periodical and spot surveys of public opinion about nuclear power generation since 1993. Spot surveys were carried out in a timely fashion to look into the impacts of the 1999 criticality accident at JCO Co., the Mihama-3 accident, and other events affecting public perceptions toward nuclear power. One of these was the accident at the Monju prototype fast breeder reactor, and another was the JCO accident, which led to increased anxieties and the sense of risk of nuclear accidents, according to public opinion surveys conducted two months after each accident. However, no statistically significant changes were observed in public opinions on the use of nuclear power. Follow-up surveys one year after each accident reported decreases in the earlier level of anxieties and sense of risk. Thus, the impact of the accidents on public opinion was temporary.

Second, this project looks into the structure of public perceptions of nuclear power generation and into public acceptance. This project has included a study on the structure of public perceptions toward radiation, research on the opinions of citizens living in areas hosting nuclear power plants, research on safety and ease of mind, research on the nuclear power plant aging degradation issue, on a nuclear fuel cycle, and on public acceptance of emerging issues such as the introduction of a probabilistic safety goal. In the coming research, the project will undertake surveys focusing on the structure of public perceptions toward radiation, radioactive waste, age-related degradation, and seismic safety, with due consideration of how the results of this research and the latest developments in the climate may affect the level of relief and reliability among the general public.

Finally, this project is committed to research projects on communications and transmission of information about nuclear power generation. In this category, we have implemented research projects on the way information should be transmitted for the purpose of raising a sense of security among the general public about nuclear power generation, communication concerning nuclear power, and dialogue activities to establish coexistence with a local community. In the coming research, we will vigorously implement surveys and research related to information service, risk communication, and coexistence with a local community, with due consideration of the results of this research and recent changes in the environment.

The Social Awareness Research Project has published two books in Japanese that describe the achievements it has accomplished through research. The first book is entitled *The Search for Peace of Mind: Human Science of Safety — Issues for the 21st Century* (INSS, 2001). It summarizes several research projects that the Social

Awareness Research Project has carried out over the years, the results of surveys and research conducted to look chiefly into safety, security, and relationships with human beings and society associated with environmental and energy issues from the perspectives of social awareness and human factors.

The second book is entitled *The Data Shows Public Attitudes on Nuclear Power: Ten Years of Ongoing Surveys* (INSS, 2004). This book represents an attempt by the Institute of Social Research's Social Awareness Research Project to focus on public perceptions of nuclear power generation, based on data amassed through ten-year-long continuous public opinion polls. During the ten-year period, events like the Monju accident, the JCO criticality accident, and the Tokyo Electric Power Company scandal occurred, raising questions about the safety and reliability of nuclear power generation. How did these accidents and other incidents impact public perceptions toward nuclear power generation? This question can only be answered by a continuous public opinion poll based on scientific methods.

The book first portrays the current public opinions. It then compares the present public opinions with those prevailing ten years ago and clarifies the extent of changes and variations in public perceptions caused by the nuclear plant accidents and incidents judging from the developments during the period. Many analysts argue that women tend to take a more negative view of nuclear power than men do. Accordingly, the book attempts to determine if there is a difference in opinion between females and males based on data, by verifying where there is a difference in perceptions between women and men and describing the extent to which women's views differ from those of men. Females tend to be less exposed to media information than males and may consequently have less knowledge about nuclear power. Females are less likely to have strongly positive opinions about nuclear power, but this does not mean they tend to have negative opinions. Rather, many of them have ambiguous or naturalistic opinions. The negative impact of the JCO accident on people's attitude to nuclear power was more prominent among females than males. However, recovery from the negative impact was faster among females than males. Thus, attitudes to nuclear power were more changeable among females than males.

One of the salient features of this book is that it analyzes how perceptions toward nuclear power are associated with various views. In relation to this question, the book discusses the relationship of public perceptions of electricity deregulation to those of nuclear power and uses multivariate analysis techniques to analyze these issues. The book provides illustrations so that readers can read it without being bothered by the details of the analytical approach. In addition, the book notes that there is a disparity between the commonly assumed public opinion of nuclear power and the actual public opinion.

Energy Issues Research Project

Over the years, the Energy Issues Research Project has been doing research into the desirable way of using energy and the impacts of human activities and energy use on the environment, focusing on nuclear power generation as one of the crucial energy sources for the modern society. In response to the growing importance of finding adequate responses to come to grips with energy and global environmental problems, the focus of research several years ago shifted to "support for promotion of energy and environmental education." The Energy Issues Research Project is currently carrying out research in the following two areas: (*i*) research on support for promotion of energy and environmental education and (*ii*) research on the diffusion of knowledge of life-styles and energy and environmental problems. In regard to the second area, the project researchers have made it a rule to focus on topics that can be reflected in school education, and promote research projects by narrowing down the focus to a few selected subjects. Some of the results are described in the following.

Research on Support for Promotion of Energy and Environmental Education

As a private research institution, INSS is not in a position to directly conduct research into school education. Accordingly, we adopted a system under which joint workshops were organized with school educators and INSS researchers, to develop a school curriculum and carry out practical research.

Workshops were organized in the Kanto area, the Kansai area, and the Fukui area where INSS is located. The Kanto area workshop aimed at proposing a curriculum and a learning method based on its educational ideas by analyzing energy and environmental education from the theoretical point of view. By contrast, workshops in the Kansai area and the Fukui area sought to come forward with proposals on specific learning experiences by actually conducting classes after consideration of the results of the Kanto area workshops.

The Kanto area workshop compiled the results of research carried out before 2004 and published a teaching aid intended to provide energy and environmental education at integrated learning classes. "Integrated learning classes" are classes in which teachers and children set themes on cross-subject topics independently and study those themes; no textbooks are used. In a sense, the teaching aid is a textbook that the workshop furnished in order to provide energy and environmental education at the integrated learning classes. The Kansai area workshop is also in the process of publishing the practices on which it has experimented. In 2007, a lower secondary school decided to promote schoolwide energy and environmental education, and its efforts were rewarded with publishing a book of the practices that summarized the path it followed. The school was also officially commended as the nation's best lower secondary school for its initiative. We are planning to take up not only school efforts of this sort but also to compile the best practices on individual themes of study and publish them as books.

Research on the Diffusion of Knowledge of Life-styles
and Energy and Environmental Problems

The following two cases can be cited:

- *Research on perceptions of people who consider that nuclear power generation is the cause of global warming.* This research was touched off by a telephone interview conducted with Japanese, American, German, and French nationals in 2001. In reply to a question asking whether nuclear power generation is a factor that accelerates global warming or a factor that checks it, answers obtained were equally split in all countries surveyed.

Accordingly, we carried out a questionnaire survey of inhabitants in the Kansai area of Japan to examine what public perceptions generate such a result. As a consequence, the survey found that people may combine a negative image of nuclear power generation with fragmentary knowledge of global warming to produce the misunderstanding that nuclear power generation is also an accelerator of global warming. In order to eliminate this misunderstanding, we must make efforts not only to capture public trust in nuclear power but also systematically to convey knowledge of nuclear power generation and global warming to the general public and help members of the public to gain a correct understanding of the issues.

- *Development of a card-based teaching material to help people take a fresh look at their lifestyles.* When we conducted a field survey of energy and environmental education in Europe, we had an opportunity to examine teaching material consisting of twenty cards portraying the scenes of lives of four families with different lifestyles. The teaching material allows children to associate the scenes described in the cards with the behaviors of their families and reconsider their lifestyles. It is simple yet easy to handle and a good teaching material that could develop children's full potential depending on leaders' abilities. However, the cards describe the scenes of European people's lives, so they cannot be used as they are in Japan. Accordingly, we are now in the process of developing a Japanese version of the teaching material and conducting classes to verify its effectiveness.

At the G8 Summit, held from July 7 to July 9, 2008 in Toyako, Hokkaido, the G8 leaders agreed to adopt the goal of achieving at least 50% reduction of global emissions by 2050 and to seek to share it with all parties to the United Nations Framework Convention on Climate Change (UNFCCC). An attempt to promote the understanding of efforts toward prevention of global warming is acquiring increasingly greater importance. Turning to the Japanese educational world, however, we notice that when the new education guidelines reflecting a review of the education with the latitude program were announced this spring, integrated learning classes that had long provided an opportunity to promote energy and environmental education in school education were reduced as from the school year 2011. We are in the process of developing teaching material for energy and environmental education that can be included in the curriculum, in order not to allow this move to disturb the flow of energy and environmental education at schools. As education intended to help people gain a correct understanding of nuclear power generation achieves greater importance, the research project is planned to be further promoted.

EFFECTS OF THE MIHAMA-3 ACCIDENT ON HUMAN FACTOR-RELATED RESEARCH

The accident at Mihama Unit 2 was attributable to underestimation of the importance of construction management and quality control (see Chapter 2). Specifically, the design significance of anti-vibration bars was not properly recognized at that time, and the quality control division failed to conduct appropriate inspections for confirming the insertion depth of those bars. As a result, the bars were improperly installed. The cause of the accident was investigated and countermeasures discussed both inside and outside of the company, as presented in the accident report, and a similar accident has not occurred since then.

However, thirteen years after the Mihama Unit 2 accident, the accident at Unit 3 occurred, revealing a weakness of the quality management system, although the cause was different. In the Unit 2 accident, the faulty insertion of the anti-vibration bars had been overlooked for a long time. Similarly, in the Unit 3 accident, no one suspected that an inspection item had been omitted from the list from the start, and the problem was not noticed for a long time. Clearly, even if a quality management system is in place, human assumptions may prevent us from noticing faulty conditions.

A review of human factor studies in the field of nuclear power generation shows that the studies have focused on themes of the times, such as issues concerning human errors made by an individual in the field, human–machine interfaces, safety culture, and interdepartmental communication. However, based on the causes of the accidents at Mihama Units 2 and 3 and our subsequent commitment to date, the INSS acknowledges with regret that our research has failed to focus on issues concerning the harmony between the quality management system and the people or organizations that administer it. Reflecting on this self-examination, the INSS has started to study the following themes.

A study on the effectiveness of a cross-check examines what kind of perspectives and allocations enable the checkers to effectively detect errors, such as omissions from documents and drawings. The examination includes verification through experiments.

Considering that in order to carry out measures to prevent recurrence of similar accidents we need a working environment that allows all of our staff to work on daily routines and individual items of study, we have launched a survey of a sense of "feeling busy" at workplaces. By measuring workers' senses of psychological and physical burdens at each workplace quantitatively and keeping track of the data obtained, we can determine how workers adapt themselves to changes in the working environment and carry out preventive measures that are needed in advance.

Moreover, in order to extend the focus of research from nuclear safety to occupational safety, we have surveyed serious accidents that have occurred in other industrial sectors and compiled a booklet containing the lessons learned from these accidents. Speaking of occupational safety, the importance of this safety was recognized as a new item entitled "securing occupational safety" that was added to the code of ethics of the Atomic Energy Society of Japan. The revision was approved in September 2007.

In addition, we have embarked on research projects on a quality management system and are also preparing a roadmap for human factors research. We have taken up a wider range of research areas than we did in research projects prior to the Mihama-3 accident.

FUTURE CHALLENGES FOR THE RESEARCH
CENTER FOR HUMAN FACTORS

One of the studies that forms the core of this center is a corporate safety culture survey and research project. This project assesses an organization's safety culture quantitatively through a questionnaire survey of staff, thus bringing to light the characteristics of the organization. The results of surveys conducted so far reveal that the organization's attitude toward occupational safety, as well as workers' immediate supervisor's attitude toward occupational safety, exerts an influence over individuals' safe behavior. In research projects carried out to date, we have given priority to diagnosing the organization and have concluded that communications between officers and their subordinates are problematic. From now on, we will also devote our efforts to coming up with measures for improvements befitting the organization.

The Research Center for Human Factors is also carrying out independent surveys of safety culture, a sense-of-busyness, and officers' leadership. By examining the results of these surveys and considering mutual relations between these and other surveys, we will be able to unify and upgrade our survey methods.

Taking advantage of the facts that the INSS is located near nuclear power plants, that it has a solid channel of communications with electric utility corporations, and that it is undertaking joint research projects by former workers at power plants and researchers who majored in psychology, we will in the future devote much greater efforts to doing site-connected human factors research.

One example of site-connected research is a safety culture field survey and research project. Under this research project, INSS researchers are embedded in a workplace to gain first-hand experience of the atmosphere at the workplace and thereby collate it with the results of a safety culture questionnaire survey, to provide precise consulting services to the site. The results of this research project have been reported at the International Symposium on Human Factors 2007, which was held by INSS in November 2007, and are also described in Chapter 10. At each workplace, the staff conducts various activities to improve the safety culture. We study the details of such improvement activities through interviews and summarize the results so that the information from each workplace can be made available to other workplaces. We believe this will help improve the entire organization in general, and we hope that the results of the project will be of help to others as well.

In closing, we hope that this book will contribute to promoting safety not only in the nuclear sector but in other industrial sectors as well.

REFERENCES

INSS. (2001). *The search for peace of mind: Human science of safety — Issues for the 21st century.* Japan: President Inc. (ISBN: 4-8334-9067-6).

INSS. (2004). *The data shows public attitudes on nuclear power: Ten years of ongoing surveys.* Japan: President Inc. (ISBN: 4-8334-9098-6).

Itoigawa, N., Fahlbruch, B. & Wilpert, B. (2004). *Emerging demands for the safety of nuclear power operations: Challenge and response.* Boca Raton, FL: CRC Press.

Misumi, J., Wilpert, B. & Miller, R. (1998). *Nuclear safety: A human factors perspective.* Boca Raton, FL: Taylor & Francis.

Wilpert, B. & Itoigawa, N. (2001). *Safety culture in nuclear power operations.* Boca Raton, FL: Taylor & Francis.

2 The Mihama-2 Accident from Today's Perspective

Masaharu Kitamura

CONTENTS

INTRODUCTION

On February 9, 1991, an accident occurred at KEPCO's Mihama nuclear power station, Unit 2, which resulted in the first-ever activation of an emergency core cooling system (ECCS) in Japan. Although a small amount of radioactive material was released to the external environment, the overall impact was negligible. However, the public media in general expressed serious concern about the possibility of nuclear disaster because they believed that the activation of the ECCS could have led to a catastrophic scenario. The actual environmental impact of the accident was unquestionably minimal. Nevertheless, the accident has, in various contexts, caused significant change to the relationship between the nuclear industry and Japanese society.

In this chapter, a brief outline of the accident will be given. Then, the effects of the accident will be discussed from a technical, organizational, regulatory, and social perspective, followed by a summary of the lessons learned from the accident and the measures taken to prevent similar accidents from recurring. Although measures taken in 1991 might have been deemed the most appropriate at the time, it would be meaningful to reconsider the effectiveness, impact, and unintended consequences of the actions taken from today's perspective. This will make it possible to acquire additional lessons from the evaluation of use for future implementation of countermeasures against different types of trouble or accidents. Based on the reconsideration, the chapter will conclude with some proposals for improving the relationship between Japanese society and its nuclear community.

OUTLINE OF THE ACCIDENT

This section chronologically outlines the history of the Mihama-2 accident, covering the causes of the accident, the discovery of threatening signs, the outbreak of the accident, the operating crew response, and the closure of the accident. The important issues to be discussed later in this chapter are also specified. A simplified schematic diagram of the Mihama-2 nuclear power plant is given in Figure 2.1. The

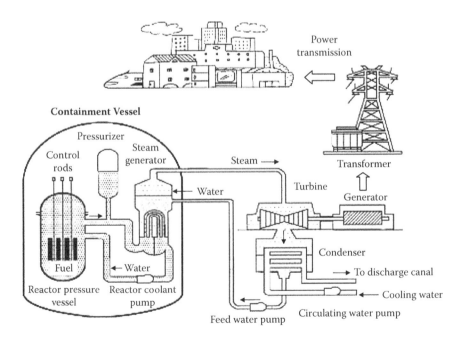

FIGURE 2.1 A simplified schematic diagram of the Mihama-2 nuclear power plant.

heat is produced by nuclear fission in the reactor pressure vessel. The nuclear fuel is confined inside the pressure vessel, which is a 40 m long cylinder of steel with semispherically shaped top and bottom domes. The pressure inside of the reactor pressure vessel is about 157 kg/cm^2 so that the cooling water is kept in a single-phase state without boiling. The temperature of the coolant at the hot leg (i.e., outlet piping from the reactor pressure vessel) is about 320°. The high-temperature non-boiling water is then transported to the steam generator by the reactor coolant pumps.

A sketch of the steam generator is given in Figure 2.2. Although the description is simple, it should be sufficient to understand the causes and effects of multiple events to

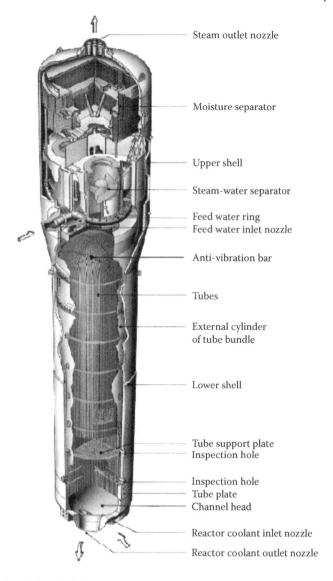

Steam outlet nozzle

Moisture separator

Upper shell

Steam-water separator

Feed water ring
Feed water inlet nozzle

Anti-vibration bar

Tubes

External cylinder
of tube bundle

Lower shell

Tube support plate
Inspection hole

Inspection hole
Tube plate
Channel head

Reactor coolant inlet nozzle

Reactor coolant outlet nozzle

FIGURE 2.2   A sketch of the steam generator.

TABLE 2.1
Timeline of Major Events in the Mihama-2 Accident

Date	Time	Action or Event
February 9, 1991	12:24	First cautionary signal was issued, reporting an increase in radiation levels on the secondary side. This was a cautionary signal distinct from ordinary alarms. Operating crews intensified their monitoring.
	12:40	It was observed that the time-sequential recording of secondary-side radiation levels demonstrated a slight rising trend. The shift supervisor, alarmed by the possibility of a primary coolant leakage to the secondary side (tube rupture), ordered a sampling and analysis of secondary coolant.
	13:20	The analysis results were obtained. Although the reactor had two trains of steam generators (A and B), the radioactivity concentration in the train A coolant was evidently higher than the radioactivity concentration in the train B coolant, which in fact was below the detection threshold. The shift supervisor requested further analysis for reconfirmation. (*Issue 3: Strategies for rapid and reliable response to troubles.*)
	13:40	Another alarm issued, reporting an increase in secondary radiation level readings. Various automated switching operations were executed, designed to prevent radioactive discharge by changing the destinations of exhausts and drains that were normally discharged into the atmosphere or the ocean.
	13:45	Another alarm issued, reporting a drop in the primary coolant system (low pressurizer pressure). This was followed, in quick succession, by a pressurizer low-water-level alarm and a steam generator high-water-level cautionary alarm. These clearly indicated that a loss-of-coolant accident had occurred.
	13:47	Operating crews decided to shut down the reactor. A procedure for manually decreasing the generator output was initiated.
	13:50	Low pressurizer pressure triggered an automatic reactor shutdown (a scram). This was followed by an automatic shutdown of turbines and generators. About 7 s after reactor shutdown, a simultaneous reporting of the low-pressurizer alarm and the pressurizer low-water-level alarm caused the emergency core cooling system (ECCS) to start up. Emergency diesel generators started up in response to the automatic shutdown of turbine-driven generators. The ECCS-related equipment began injecting water into the primary coolant system.

TABLE 2.1 (CONTINUED)
Timeline of Major Events in the Mihama-2 Accident

Date	Time	Action or Event
	13:55	To isolate the damaged steam generator A, operating crews closed the main steam isolation valve. However, it was not possible to confirm complete closure of the main steam isolation valve from the main control room. Therefore, workers had to rush to the site and close the valve manually. (*Issue 4: Assuring Maintenance Quality and Improving Maintenance Procedure Manuals.*)
	14:10–14:25	Operating crews repeatedly attempted to open the two pressurizer relief valves to lower the primary coolant system pressure, with the aim of decreasing the outflow of primary coolant to the secondary side. However, neither of the two valves would open. (*Issue 5: Improving the working environment to optimize maintenance activity.*)
	14:34	Having decided to give up trying to open the pressurizer relief valves, operating crews initiated the alternative procedure of starting up the pressurizer auxiliary spray to lower the primary coolant system pressure. (*Issue 6: Ensuring the thoroughness of operation manuals.*)
	14:48	Having confirmed a sufficient drop in the primary coolant pressure, operating crews terminated the depressurization procedure by closing the pressurizer auxiliary spray valve.
	15:55	A procedure for increasing the boron concentration was initiated to prepare for a shift to cold shutdown. (*Increasing the boron concentration of the cooling water increases the absorption of neutrons, hence lowers the reactivity.*)
	16:18	The procedure for increasing the boron concentration was completed.
	16:40	Operating crews opened the turbine bypass valve to accelerate the cooling of the primary coolant system using the condenser.
February 10, 1991	02:37	The reactor completed a shift to cold shutdown.

be described later in this section. The function of the steam generator is to transfer the heat from the high-temperature water flowing through the primary coolant loop (consisting of reactor vessel, coolant piping, primary side of the steam generator, and coolant pump) to the secondary side of the steam generator. Because the pressure of the secondary side is lower than the primary side, the secondary side water causes significant boiling and thereby provides steam to the turbine, which in turn drives the generator to produce electricity. In the loop between the reactor pressure vessel and the steam generator, a pressure control tank called a pressurizer is installed. The purpose of the pressurizer is to maintain the pressure in the primary loop at a fixed value. Because the pressurizer is connected to the primary loop, the lower part of the pressurizer contains non-boiling water. The upper part, however, contains steam. When the pressure in the primary coolant loop is lower than the specified value, an electrical heater in the lower part is turned on to produce steam so that the pressure is again increased. When the pressure is higher than the specified value, cooling water is sprayed into the steam portion of the pressurizer leading to condensation of the steam and thereby to a reduction of the pressure. If the pressure despite that still is high, a pressure relief valve opens automatically to release the steam and thereby reduce the pressure.

The steam generator that caused the accident was constructed in 1970. The construction was undertaken by Mitsubishi Heavy Industries, Ltd., under the guidance of Westinghouse Electric Corp., USA. The post-accident inspection revealed that the immediate cause of the accident was the imperfect insertion of anti-vibration bars for a U-tube bundle inside the steam generator (Figure 2.3). It remains unclear why such imperfect workmanship was allowed to happen. What is evident, however, is the absence of countermeasures to prevent or subsequently detect such a gross violation. (*Issue 1: Quality control in the design and construction stages.*)

Rapid flows exist inside and outside the steam generator tubes, which creates an environment where various vibrations are produced. Because the U-tube bundle did not have proper anti-vibration measures installed, it was subject to intense vibrations at levels that

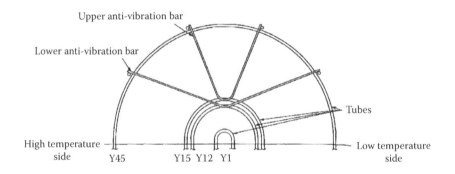

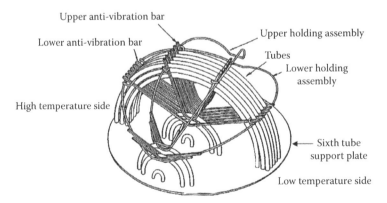

FIGURE 2.3 Diagram showing anti-vibration bars for a U-tube bundle inside the steam generator.

surpassed design expectations. Such vibrations accelerated the metal fatigue of steam generator tubes, resulting in eventual major damage. Adding to the mystery of the imperfect workmanship was the inability to detect the problem during the acceptance inspection performed upon completion of work as well as during the number of annual outages in subsequent years. (*Issue 2: Quality control in acceptance inspection stage.*)

Whatever the reason, the imperfect workmanship from the construction stage remained unnoticed. In July 1972, the Mihama-2 reactor entered into commercial operation. During the eighteen years since then, the steam generator, burdened with this (unknown) defect, seemed to continue its operations without any problems. However, over the years, the steam generator tubes slowly but steadily degraded from metal fatigue induced by the vibrations.

Table 2.1 provides an overview of the course of the events of the accident. Even though the course of events was more complex in reality, the description of the accident scenario has been simplified to put emphasis on the issues to be discussed later in the chapter.

A small amount of radioactive material was released to the environment during the accident, although a description of that has been omitted. According to nuclear experts, the amount released was far below safety hazard levels. Nevertheless, it goes without saying that any radioactive release is undesirable.

LESSONS LEARNED FROM THE ACCIDENT AND MEASURES IMPLEMENTED TO PREVENT RECURRENCE

First, this section discusses the lessons learned from the Mihama-2 accident and the countermeasures that were implemented, mainly from the viewpoint of their adequacy and rationality in consideration of the situation *at the time*. In the following section, these countermeasures are discussed in the light of the knowledge that is available today. Although this may tell us what should have been done from a present perspective, it is not intended as undue criticism of what actually was done. It is hardly surprising that responses that were appropriate at one time may seem insufficient at a later time. It is nevertheless not the purpose of this section to provide a comprehensive review on all the lessons learned and all the countermeasures implemented. The following discussions are focused on topics that deal with the important issues that concern human factor issues at large.

More precisely, the issues addressed in the outline of the accident will be discussed from the perspectives of human factors at the personal level, organizational level, and social level, which today has an even larger significance. In this section, the discussions will be focused on the personal level and organizational human factors. The concept of human factors at the social level will be presented in the following section, with the emphasis on how managing the nuclear industry–society relationship may influence human factor issues at industrial sites. Human factors at the society level in many ways correspond to what conventionally has been referred to as performance-shaping factors (PSFs), including such factors as workplace culture, workers' morale, and corporate policies (Swain & Guttmann, 1983). However, this chapter introduces the concept of human factors at the social level, based on the observation that these PSFs in reality are the products of relationship management between the organization and the society. The efforts toward the improvement of these PSFs will therefore not go beyond a certain threshold, if pursued only within the organization.

ISSUES 1 AND 2: QUALITY CONTROL IN THE DESIGN AND CONSTRUCTION STAGES AND DURING THE ACCEPTANCE INSPECTION

Lessons learned: A lack of awareness about the importance of anti-vibration bars in the context of plant design possibly led to imperfect workmanship. Even though workers must have been aware that the anti-vibration bars had to be installed in order to serve a certain purpose, they were probably unaware that an imperfect installation could cause a tube rupture accident in the future.

The fact that the quality control department did not inspect the insertion depth of the anti-vibration bars also suggests a lack of attention to construction management and quality control. Thus, the site workers must have reported the work to be complete even though the anti-vibration bars had not been inserted as deeply as intended, and the supervisors must have approved the work without conducting a proper inspection.

Measures to prevent recurrence: To prevent a recurrence of these events, the regulatory authorities included anti-vibration bars in the list of components to be covered by the construction plan approval and the pre-service inspection. In addition, a requirement was added to ensure that the insertion and installation of anti-vibration bars be checked during annual outages.

In the context of voluntary measures, electric power companies decided to reinforce quality assurance for the design, manufacturing, construction and operation stages. For design, manufacturing, and construction quality assurance, electric power companies emphasized stronger strategies for strengthening procurement control.

ISSUE 3: STRATEGIES FOR RAPID AND RELIABLE RESPONSE TO TROUBLES

Lessons learned: It is not realistic to expect an immediate decision for a major action like reactor shutdown following reports of minor signs of abnormality by operating crews. However, even after the first water quality analysis had clearly demonstrated an increase in radioactivity concentration, the shift supervisor requested further analysis for reconfirmation. T   validity of this request might be viewed as questionable although not unrealistic. If action had been taken immediately to lower the reactor output, it is likely that the accident scenario would have developed along lines that allowed easier responses, even though it might not have been able to prevent the steam generator tube from rupturing in any case. Enhancing the instrumentation system was thought to be desirable for supporting quick decision making by operating crews.

Measures to prevent recurrence: A new leakage monitoring method was introduced for detecting any leakage to the secondary system without performing a time-consuming water quality analysis. This method is based on the use of the N-16 monitor that monitors the secondary coolant radiation levels. Electric utilities have prepared manuals concerning this method.

ISSUE 4: ASSURING MAINTENANCE QUALITY AND IMPROVING MAINTENANCE PROCEDURE MANUALS

Lessons learned: Operating crews encountered some difficulty in closing the main steam isolation valve, which demonstrated the importance of maintenance quality assurance. Adjustment of the surface roughness on sliding portions in the valve was insufficient during maintenance activities, which resulted in larger friction and difficulty in operating the valve.

Measures to prevent recurrence: Maintenance procedure manuals were revised to prevent such cases of imperfect maintenance from recurring. In addition, a valve closure supporting device was introduced to facilitate the closing of a valve that resists closing.

ISSUE 5: IMPROVING THE WORKPLACE ENVIRONMENT TO OPTIMIZE MAINTENANCE ACTIVITY

Lessons learned: The pressurizer relief valves failed to open because a valve in the piping that supplied the air needed to actuate these valves had been closed by mistake. The direct cause of the valve opening failure was therefore a human error committed by maintenance personnel. A detailed analysis, however, revealed a design flaw. This particular air supply valve was assigned multiple functions, which made it difficult for the site workers to understand its purpose.

Measures to prevent recurrence: As a result of the identified problems, the hand-operated valves that acted on components with important safety functions were modified to significantly reduce the probability of erroneous operation. Specific modifications included locking control, valve position display, and a clearer description of inspection targets in inspection manuals.

ISSUE 6: ENSURING THE THOROUGHNESS OF OPERATION MANUALS

Lessons learned: In the face of an emergency caused by the steam generator tube rupture, the operating crews unexpectedly encountered functional failures of important safety components; namely, the imperfect closure of the main steam isolation valve and the pressurizer relief valves failing to open. That they were able to handle the accident by responding creatively to an unexpected situation should be positively acknowledged. However, the procedures chosen by the operating crews were not mentioned in manuals.

Measures to prevent recurrence: In view of the above, the procedures chosen by the operating crews in response to the accident were clearly documented and included in the operation manuals.

The previous discussion summarized the lessons and measures to prevent recurrence in terms of the important issues to be discussed further in this chapter. In addition to the recurrence prevention measures mentioned, the regulatory authorities reinforced their guidance and supervision over maintenance activities. This action pertained to all six issues.

RETROSPECTIVE EVALUATION FROM A CURRENT PERSPECTIVE

All the measures taken to prevent a recurrence of the events described in the previous section must be regarded as appropriate in the light of findings that were available at the time. However, practical knowledge on nuclear safety has evolved much since then. This section, therefore, attempts a retrospective evaluation of the measures to prevent recurrence that were adopted in 1991, from a current perspective.

A critical reevaluation of the judgments and decisions of seventeen years ago from today's perspective may appear too harsh in ordinary cases. However, nuclear installations operate over a long period of time, and the operators of such installations are required to ensure long-term safety. We should again recognize the importance of the fact that a latent failure produced in 1970 culminated into an accident in 1991. As a matter of fact, the organization that produced the latent failure is the most responsible part. In addition to this, another cause of the accident was, strictly speaking, the long period during which all employees, who might have been alarmed by the possibility that the plant had not been constructed exactly according to the design plans, failed to attempt any kind of investigation. In other words, they uncritically accepted the status quo and never thought to question its premises. In this context, from today's perspective, it should be worthwhile to look back and weigh the measures to prevent recurrence relative to the problems of the past that appeared to have been settled. A similar retrospective analysis, with emphasis on procedural complexity in recovery operations, was conducted by Hollnagel (1998) for the tube rupture accident at Ginna nuclear power plant, which occurred in 1982. The present retrospective analysis is not specific to post-accident recovery but puts more emphasis on broader perspectives such as quality control in design and construction stages, quality control in maintenance activities, and management of social relationships between the nuclear industry and society. This reevaluation is attempted below for each of the important issues identified in the previous section.

ISSUES 1 AND 2: QUALITY CONTROL IN THE DESIGN, MANUFACTURING, AND CONSTRUCTION STAGES AND DURING THE ACCEPTANCE INSPECTION

Including descriptions of the installation of anti-vibration bars in construction plans and pre-service inspection documents is reasonable as a measure to prevent recurrence, considering the event that took place in the past. However, deeper thinking will reveal that the *essential cause* of the Mihama-2 accident was *that a situation was allowed to develop that went beyond common sense* and that it furthermore *was overlooked for many years because it had escaped everyone's attention*. Trying to imagine what is unimaginable according to common sense is naturally a difficult challenge. The quality assurance practices for design, manufacturing, and construction introduced in 1991 as a measure to prevent recurrence are reasonable to a certain degree as an approach to this challenge. However, merely practicing quality assurance by following basic instructions in documented guidelines is unlikely to solve essential problems. Improving the real effectiveness of quality assurance activities will require a proper safety culture (International Nuclear Safety Advisory Group

[INSAG], 1991, 1999); this includes forming a questioning attitude, as emphasized by the International Atomic Energy Agency (IAEA), and the sharing of engineering ethics and traditional craftsmanship. The introduction and establishment of such fundamental measures requires the firm determination and sustained commitment not only of the field engineers but also of top management.

It is commonly agreed among human factor experts that errors at the workplace should not be attributed only to those who committed them. As a matter of fact, it is necessary to create a workplace environment that tries to increase beneficial performance-shaping factors and reduce error-forcing contexts (Cooper et al., 1996). In this regard, top management has to pay attention to many factors, including employment and assignment, education and training, better treatment, and proper distribution of relevant resources. This has to be clearly recognized and practiced by top management. On the other hand, there will be engineers who execute their tasks reliably with pride even in an organization that does not entirely fulfill such conditions, and there is a strong demand for such engineers. Human factor studies and their practical application to workplaces should always take into account the interactions between top management and field engineers concerning these apparently contradictory demands.

Issue 3: Ensuring Rapid Response to Troubles

As far as this accident is concerned, the conflict between the need for an accurate judgment and the need for rapid action to prevent an accident seems to have been solved by a decision to strengthen the monitoring system. However, this essentially is the traditional problem of trade-off between avoiding false alarms and missing true alarms following detection of signs of abnormality. Therefore, the problem of conflict in a general context still remains. It is desirable that ambiguities concerning judgment criteria, in the face of conflict in general, are minimized by the establishment of proactive guidelines. Maximum efforts should be made to present the guidelines as much as possible in the form of explicit statements.

Nevertheless, however complete the rules may be, it is impossible to proactively address all possible cases. Therefore, the ultimate criterion is often defined using an abstract term such as "the highest priority on safety." However, it should be noted that putting the highest priority on safety does not mean: "Shut down the reactor after any sign of trouble that cannot be corrected immediately." Shutting down the reactor, in itself, produces a major transient to the nuclear power plant, causing many consequences. Furthermore, it produces a large burden on operating crews. A reactor shutdown is not a procedure to be chosen unconditionally or indiscriminately. It is necessary to have a firm comprehension of the set of potential event scenarios and to evaluate in advance the probability and severity of the consequences of each scenario. This methodology for risk-based decision making should be emphasized even more than it currently is.

With regard to the decision-making guidelines discussed previously, another important challenge for both regulation authorities and society is to have a shared understanding of the basic policies behind such guidelines. This will be discussed in the next section.

Issue 4: Maintenance Quality Assurance

Maintenance procedure manuals were revised to prevent the recurrence of the imperfect maintenance of the main steam isolation valve. Deep reflection into the causes of this problem should not end by recognizing imperfection in specific procedure manuals. Even in the absence of explicit instructions on specific activities in procedure manuals, workers may make their own choices when executing specific tasks. In this case, the workers made a choice about the surface roughness of sliding components in the valve. Making decisions on one's own without clearly established reasons and in the absence of explicit instructions is a pattern of behavior that needs to be understood and corrected.

In this context, it is not sufficient that workers only know *how* they should execute their tasks; it is expected that they also know *why*. It is further expected that there are senior engineers who also know *what if*; that is to say, people who have an ability to foresee the consequences of their choice of action.

Of course, in reality, workers cannot be expected to reflect upon know-why and what-if at each step of a work procedure. However, aspiring toward such knowledge could be supported better in the context of education, training, and personal growth programs. Particularly in the education and training activities, it is strongly advised to incorporate know-why and what-if drills in addition to know-how instructions. If the trainees are educated by means of know-why and what-if drills, they will gain much higher competency in conducting their own assigned activities than by blindly memorizing know-how manuals.

Issue 5: Improving the Workplace Environment to Optimize Maintenance Activity

This topic addresses not only the correction of the instrumentation system complexity that caused an operation error but also more general improvements. Therefore, any of the problems at hand were actually solved. However, as time passes, various components of improved design are introduced to plants and various modifications are implemented to accommodate them. Through this process, some empirical knowledge from the past can be forgotten. For example, there was a case in which a hand-operated valve was replaced by a motor-operated valve to prevent human-factor errors. While this successfully eliminated valve operation errors, a change in the vibration characteristic of the piping resulting from this replacement ruptured the piping by fatigue.

With plants operating over several decades, the number of people who have worked since the plant was constructed is rapidly diminishing. In the future, the challenge of passing on practical knowledge and experience from the past to the present will have to be seriously considered.

Issue 6: Ensuring the Thoroughness of Operation Manuals

The improvements required under this topic are basically similar to the improvements suggested for Issue 3: maximum formulation in the form of explicit statements and the provision of guiding principles about responses to unformulated events. However, with this topic, I intend to comment on another question concerning guiding principles.

The question is this: *Should an operator never undertake an action that has not been prescribed by manuals, or should an operator choose an action that is evidently better for improving the situation rather than following manuals?* Because this is an ultimate dilemma, it is desirable that an appropriate solution be found before someone is really faced with a situation involving such a difficult choice. However, we cannot deny the possibility that such a situation may emerge and suddenly put someone in this dilemma without any preparation time.

No simple answer to this problem exists. However, the frequent repetitions of what-if drills described earlier (Issue 4) may help to develop an ability to deal with a dilemma like this. It appears that drills like these have already been introduced to a certain degree, but insufficiently so. As an immediate benefit, one may expect that these drills will lead us to discover the shortcomings of manuals and procedures. An even greater benefit from these drills that can be expected is a deeper awareness of problems and a better understanding by operating crews.

This section has emphasized that it was necessary to take actions from a wider perspective based on general implications, on top of measures to prevent recurrence of specific cases of functional failures and operation errors identified from the analysis of event scenarios experienced in the past. It is only through such actions that it becomes possible to reach a higher level of safety, which is far superior to the "tombstone safety" that is built on reactive measures against specific failings in the past.

In the previous sections concerning each of the important issues for discussion, I emphasized challenges that deeply concern human factors at the personal and organizational levels, particularly regarding Issues 1 and 2. Researchers often distinguish between personal human factors and organizational human factors for the sake of research. However, this distinction is meaningless when seeking practical applications. Those who seek to introduce the results of research into workplaces are expected to take note of this issue as they make efforts to improve the working environment at large in pursuit of higher safety.

DISCUSSIONS ON RISK FACTORS THAT INTERPLAY WITH SOCIETY

This author believes that further research efforts must be introduced based on a consideration of social factors. It is widely recognized that a considerable portion of nuclear risk is related to human factors. However, it is less recognized that the human factors have significant interactions with regulatory *and* social factors. An issue of particular importance in this context is the need for normative consensus on the concept of nuclear safety between the nuclear industry and society.

One typical example of the need for sharing the concepts of safety with the society was already addressed in relation to Issue 3. Even though plant workers will understand that placing the highest priority on safety does not mean the shutdown of the reactor at any sign of trouble, the public may think differently. Unless this recognition is shared with the society, this mismatch in understanding may frequently cause serious social problems.

Another example concerns prioritization of actions for managing an accident. When a major failure or accident has happened, it may be a reasonable decision from the engineering point of view to postpone responses to smaller troubles with less important components that virtually do not present much risk, considering that the safety of the nuclear power plant as a whole is more important and that the number of plant workers is limited. In a case like this, one may choose to deviate from operation manuals, which is a question discussed in relation to Issue 6. In the face of a number of tasks to be carried out, the assignment of priorities to different tasks and the optimum distribution of resources in view of these priorities are issues that require fundamental discussions. However, deviations from operation manuals are likely to be blamed by the public unless the necessity of prioritization is understood in advance.

As the third example, there is a need for nationwide discussions and the formation of consensus as to the question of how seriously one should take the release of a minute amount of radioactive materials.

The release of a minute amount of radioactive materials is evidently a failure because such a release is essentially undesirable. However, in the case of the Mihama-2 accident, we may think that the operating crews very successfully mitigated the accident with limited resources and in spite of difficulties, as described in relation to Issues 4–6. Further discussions are needed as to how nuclear experts and society should evaluate such cases with conflicting normative requirements.

This challenge — revisiting normative requirements being treated as obvious — suggests the need to establish a new kind of relationship between the nuclear industry and society. I would like to address such needs in general by presenting in the following section another important issue of discussion that has not been treated in an explicit manner since the accident.

Issue 7: Normative Principles of Nuclear Safety to be Shared with Society

The implication of Issue 7 can be better understood by considering the current situation of the interactions between the nuclear industry and society following an accident.

Reactive actions have dominated in the past — after every accident or trouble that produced concerns about nuclear safety, utilities and regulatory authorities have expressed their apology to the public for their failings and then introduced new countermeasures.

The post-accident regulatory requirements can impose extra workload and time pressure as a result of "improved" regulatory procedure, including higher quality assurance. The workload and time pressure can be amplified because of "improved" requirements of anomaly reporting to local governments. Furthermore, the social environment surrounding the nuclear personnel can be modified by media messages, which are mostly critical and sceptical toward nuclear organizations. The plant personnel cannot remain unaffected by a social environment that is often hostile to them. In the long run, negative outcomes become more likely for several reasons: morale of the plant personnel might be degraded, the quality of each operation for power generation and machine maintenance might deteriorate, and recruiting of human resources might also become difficult.

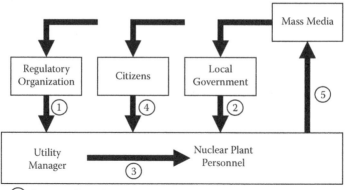

1. "Improved" regulatory requirements
2. "Improved" anomaly reporting rule
3. Additional requirements to plant personnel
4. Social atmosphere against nuclear industry
5. Trouble and anomaly information

FIGURE 2.4 The scheme of social interactions to influence on the nuclear risk.

All these effects, illustrated in Figure 2.4, can actually have significant effects for various human factors, and thus for risks in nuclear facilities. The human factors introduced through those social and regulatory channels can be named socially constructed human factors. A systematic approach to safety improvements based on the situation recognition of Figure 2.4 with sufficient consideration of the socially constructed human factors is necessary to integrate and empower the individual R&D projects.

This author acknowledges the need for such a mechanism for reactive actions, but finds it necessary to question whether such reactive actions are the best that can be done. In reference to topics that were compiled as Issue 7, with specific instantiations as Issues 3 and 6, for example, higher safety in the nuclear industry inevitably requires us to question the reasonableness of present norms. This process of requisitioning requires not only the involvement of the nuclear community but also communications with various stakeholders in society (Andersson, 2006; Andersson et al., 2006). We should not overlook that social factors are producing real impacts on risks presented by nuclear installations.

One important option for attaining the goal is certainly to improve the public relationship management between the nuclear industry and the public, for instance by implementing a new scheme of public communication (Yagi, Takahashi, & Kitamura, 2003, 2007). Although the topic is somewhat out of the scope of this book, I would like to stress that public engagement in risk identification and management is a key issue to maintaining the social relationship between experts and society in many domains of technology. Intensive efforts are necessary to improve bilateral communication, instead of the traditional public acceptance, to convince the public by providing explanatory information. Reconstruction of novel relationships and shared norms is a crucial factor for better utilization of nuclear and other technologies in our risk-conscious society.

CONCLUSIONS

In this chapter, several important issues were first extracted from the analysis of the Mihama-2 accident. Then, the lessons learned and corresponding efforts were summarized in conjunction with the issues. Unquestionably, a significant amount of effort has already been made on research and development to satisfy the requirements derived from the lessons. However, the requirements related to human factors and regulatory issues are more difficult to accomplish than the requirements related to mechanical systems and equipment. Supplementary requirements have been devised in this chapter through a reevaluation of these issues and efforts from the current perspective. The reevaluation is motivated by the aim to obtain a deeper insight into the accident, so that one can develop better countermeasures to realize a higher level of nuclear safety. Several countermeasures from the reevaluation, including the development of managing the social relationship between the nuclear industry and the public, have been proposed.

Other accidents and troubles subsequently experienced at Japanese nuclear facilities, including the fatal accident at JCO in 1999 (described in Chapter 8), the data falsification of Tokyo Electric Power Company uncovered in 2002 (described in Chapter 8), and the fatal tube rupture accident of the Mihama-3 plant in 2004 (described in this chapter), were mainly caused by defects in human factor considerations by the responsible organizations.

I do not intend to say that these accidents could have been avoided by adopting the countermeasures proposed in the preceding sections. However, the improvement of social relationship management is certainly desirable, not only for reducing the possibility of accidents, but also for post-accident social relationship recovery. The importance of social relationship management must be clearly recognized as a foundation for a higher level of nuclear safety, because public opinion is most crucial in our democratic society. The importance and necessity of an effective methodology for public communication must also be understood by the people involved in the nuclear industry.

REFERENCES

Andersson, K. (Ed.) (2006). *Proceedings of VALDOR* (VALues in Decisions On Risk), May 14–16, Stockholm, Sweden. (http://www.congrex.com/valdor2006/)

Andersson, K., Drottz-Sjöberg, B-M., Espejo, R., Fleming, P. A. & Wene, C-O. (2006). Models of transparency and accountability in the biotech age. *Bulletin of Science Technology Society, 26,* 46–56.

Cooper, S. E., Ramey-Smith, A. M., Wreathall, J., Parry, G. W., Bley, D. C., Luckas, W. J., Taylor, J. H. & Barriere, M. T. (1996). *A technique for human error analysis* (ATHEANA), NUREG/CR-6350. Washington, DC: Nuclear Regulatory Commission.

Hollnagel, E. (1998). *Cognitive reliability and error analysis method (CREAM)*. New York: Elsevier Science.

International Nuclear Safety Advisory Group (1991). *Safety culture* (INSAG-4). Vienna: IAEA.

International Nuclear Safety Advisory Group (1999). *Measurement of operational safety in nuclear power plants* (INSAG-13). Vienna: IAEA.

Swain, A. D. & Guttmann, H. E. (1983). *Handbook of human error reliability analysis with emphasis on nuclear power plant applications*, NUREG/CR-1278. Washington, DC: Nuclear Regulatory Commission.

Yagi, E., Takahashi, M. & Kitamura, M. (2003). Toward bridging a gap between experts and local community through repetitive dialogue forum. *Proceedings of VALDOR 2003*, Stockholm, pp. 419–425.

Yagi, E., Takahashi, M. & Kitamura, M. (2007). A proposal of new nuclear communication scheme based on qualitative research. *Japanese Journal of Atomic Energy Society of Japan*, 6(4), 444–459.

3 Extending the Scope of the Human Factor

Erik Hollnagel

CONTENTS

INTRODUCTION

In the history of system safety, the human factor has been considered in three quite different roles. In the first role, the human factor was looked at as a bottleneck and as something that hindered the full use of the technological potential. In the second role, the human factor was looked at as a liability or a threat, and humans were seen as something that limited performance as well as being a source of risk and failure. In the third role, the human factor is looked at as an asset, and humans are seen as a *sine qua non* for system safety.

Human factors engineering became recognized as a separate discipline around 1945. One strong motivation was the problems in the design and use of aircraft during World War II, where increased technological requirements and possibilities challenged human capabilities. Interest in the design of artifacts for human use is, however, much older. According to Marmaras et al. (1999), evidence can be found as far back as in ancient Greece twenty-five centuries ago. In our time, the first known use of the term ergonomics is by Jastrzebowski (1857), eighty years before the first issue of the French journal *Le Travail Humain* was published in 1937. One may therefore rightly wonder why human factors did not become an issue — let alone a scientific discipline — much earlier. The answer can be found by looking at the kind of technology that had been used throughout the centuries, specifically the change that took place after the industrial revolution. Although the first steam engines were taken into use around 1710, the industrial revolution is usually dated to around 1769, which was the year when James Watt patented his improved steam engine. From a modern perspective, the human use of technology — as a tool of work — was, in the centuries before the industrial revolution, characterized by the following:

- Systems were limited in size as well as in the number of parts or components. This was mainly due to the manufacturing technology and the materials available.
- Technology was uncomplicated and easy to understand. Using a modern term, technology was linear, i.e., cause–effect relations were direct rather than indirect and the effects were proportional to the causes.
- Work was mainly manual. The sources of power (a.k.a. *prime movers*) were few and simple, such as running water, wind, and husbandry. More important, the prime movers were somewhat difficult to control or, in the case of wind, completely unpredictable, and could therefore rarely be relied on to produce a constant power output.
- Artifacts were few and usually only loosely coupled. Most artifacts that were used for work were made by hand and in small quantities. This was another reason why systems were limited in size and complexity.
- System integration was limited, both horizontally and vertically. There was no way of coupling subsystems and components as we know today; that is, no way of transmitting information rapidly and effectively among components or between humans and technology. Until electricity came into common use, signals could be transmitted only by mechanical means (e.g., wires, vacuum, visible or audible signals).

Because of this, there was a "natural" fit between humans and technology, and the pace of work was set either by human capabilities or by the scale of the mechanisms. In cases where prime movers were used as a source of power (water, wind), the processes they were used for were themselves simple and limited by the mechanical characteristics of the artifacts. "Processes" were slow and therefore relatively easy to control.

All that changed dramatically with the industrial revolution, which effectively replaced the traditional prime movers (wind, water, and husbandry) with a new source of power, the steam engine. This had two crucial advantages. First, it exceeded the natural prime movers by orders of magnitude. (An illustration of that is the use of *horsepower* as a unit of power for the steam engines.) Second, the source of power could be controlled, and therefore could be relied upon to produce a constant power output. The steam engine provided a regular and reliable force that could power almost any kind of common machinery. It was, however, also a power source that was quite large and quite expensive, which meant that work had to be brought to the source of power rather than the other way around. One consequence of that was that work changed from being paced by the individual to being paced by technology. Because of the dependence on the power source, technological efficiency (machine requirements) came to dictate the design of the workplace.

The industrial revolution also introduced the industrial accident or work accidents. People have, of course, always had accidents as a part of work (which often took place at home), during major building works, when traveling on land or at sea, and of course during wars. But until the first decades of the nineteenth century, risks were accepted as more or less natural in the sense that they were directly associated with human activity rather than with failures of systems or equipment. The industrial revolution changed this by making machines an important part of work, both stationary and moving. Stationary machines, such as low-pressure steam engines, could explode (e.g., Leveson, 1994), and moving machines could hit people. On September 15, 1830, William Huskisson became the first victim of a train accident. The occasion was the opening of the Liverpool and Manchester Railway and the train was George Stephenson's Rocket. More accidents soon followed, involving exploding boilers, derailing, head-on collisions, collapsing bridges, and so on. (As an aside, the first recorded automobile death took place in Ireland on August 31, 1869, when a woman, Mary Ward, was thrown from and fell under the wheels of an experimental steam car built by her cousins. In 2002, road traffic accidents worldwide were estimated to kill 1.2 million people, with at least 20 million people being injured or disabled.) But accidents were seen as more or less inevitable consequences of using imperfect technology, and there was therefore no need to consider the human factor. (This bias toward technology is underlined by the observation made by Hale, 1978, that investigators in the late nineteenth century were only interested in having accidents with technical causes reported, because other accidents could not reasonably be prevented.)

FROM THE INDUSTRIAL REVOLUTION TO INFORMATION TECHNOLOGY

By the beginning of the twentieth century, technological developments had brought about a proliferation of sources of power, as well as new sources far more flexible than the steam engine. Prime among those were the internal combustion engine and the fractional horsepower motor (Wiener, 1993). The result was that many more processes could make use of a controllable source of power, that many more machines could be built to become part of the work, and that systems became larger and more complex. One result of that was that work became more specialized and thus completely new types of work began to appear. The pacing by technology, which had started with the steam engine and the industrial revolution, also continued. This led to the development of large and complicated processes (think of the automobile assembly lines) that necessitated specialized training of humans and specialized work analysis. The strongest example of that was the discipline of scientific management, also known as Taylorism. The core ideas of the theory were developed by Frederick Winslow Taylor in the 1880s and 1890s and were fully described in Taylor (1911). The basic principles of scientific management were to analyze tasks to determine the most efficient performance and then select people to achieve the best possible match between task requirements and capabilities.

The break with the linear developments that were started by the industrial revolution came as a consequence of the introduction of information technology. This is often itself called a revolution, the *information technology revolution*, and was marked by a number of significant inventions or creations, such as the digital computer (ENIAC in 1945), the mathematical theory of communication (1949), the formulation of cybernetics (1948), the transistor (1947–48), and the integrated circuit some years later (1958). With the creation of the basis for information technology, the development of machine capabilities — hence the demands of humans to control them — quickly exceeded what humans could naturally do. This created what is now known as the demand–capacity gap and led to the development of human factors engineering.

THE HUMAN FACTOR AS A BOTTLENECK

After the information technology revolution, and in particular its migration from the military to society at large, human factors became an issue because humans were seen as too imprecise, variable, and slow. Human capacity limitations, in performance and control, therefore meant that system performance and system productivity were below what the technology made possible. The three main solutions that human factors engineering developed to overcome these limitations were training, design, and automation. Training, supported by selection, was used to bridge the gap between what people in general were able to do and the skills, knowledge, and proficiency required to work effectively with the machines or technology. Design was used to ensure a good fit between the system and the users, epitomized by the human–machine interface. (Today this is very much an issue of display and interaction design, but until the mid-1980s most human–machine interfaces were based on more conventional knobs-and-dials technology.) Designs, however, also covered other issues, such as ease of use, comfort, productivity, safety, and aesthetics. Finally, automation used technology to overcome the problems created *by* technology. This created an uncomfortable situation that remains to this day. The problem that automation tried to solve was clearly stated by one of the first researchers who looked at the issue (Fitts, 1951): "We begin with a brief analysis of the essential functions. ... We then consider the basic question: Which of these functions should be performed by human operators and which by machine elements?"

One important product of this whole development was that task analysis became an almost universal technique to structure the human–machine interaction (e.g., Miller, 1953). This made it natural to think of systems and events as being composed of discrete and identifiable components.

THE HUMAN FACTOR AS A LIABILITY

The second role of the human factor was introduced rather abruptly by the accident at the Three Mile Island (TMI) nuclear power plant on March 28, 1979. Before the accident happened, the established risk analysis methods, such as HAZOP (hazard

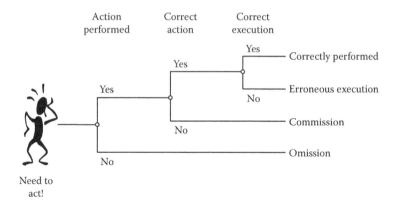

FIGURE 3.1   Errors of omission and commission.

and operability studies), FMEA (failure modes and effects analysis), fault trees, and event trees, had been considered sufficient to establish the safety of nuclear installations. After TMI, it was realized that something was missing from these methods; namely, the human factor. Even from the first descriptions of the TMI accident, it was clear that the actions of the operators played a significant role in how the events developed and hence the final outcome. This pointed to the necessity of including the effects of human actions, and more particularly the effects of incorrect human actions, in risk assessment and safety analyses.

The first reaction was to treat incorrectly performed operator actions as "human errors," in analogy with the malfunctioning of a technological system or component (see Figure 3.1). The categorization was initially limited to simple classifications of human error, such as error of omission and error of commission (Hollnagel, 2000), defined as follows:

- An error of omission is the failure to carry out some of the actions necessary to achieve a desired goal.
- An error of commission is carrying out an unrelated action that prevents the achievement of the goal.

In 1975, a committee of specialists under Professor Norman Rasmussen produced the Reactor Safety Study, or WAHS-1400 report, for the USNRC. This report considered the course of events that might arise during a serious accident at a large light water reactor, using a fault tree/event tree approach. The WASH-1400 study effectively established probabilistic risk assessment (PRA) as the standard approach in the industry for how to deal with the questions of safety and reliability of technical systems. It was therefore the natural starting point when the human factor needed to be addressed. The incorporation of human factors concerns in PRA led to the development of human reliability assessment (HRA), which first was just an extension of existing methods to consider human errors in the same way as technical failures and malfunctions, soon to be followed by the development of more specialized approaches. The details of this development have been described in several places (e.g., Hollnagel, 1998, and Kirwan,

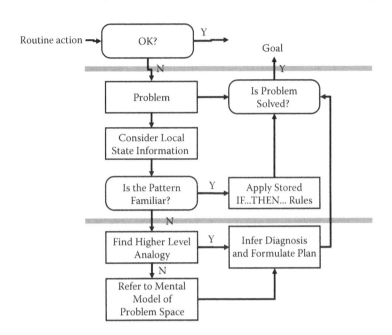

FIGURE 3.2   The GEMS model.

1994), but the essence is that human reliability became a necessary complement to system reliability — or rather that reliability engineering was extended to cover both the technological and human factors. HRA quickly became established as the standard analysis for nuclear power plant safety, although there have never been any fully standardized methods (e.g., Dougherty, 1990), or even a reasonable agreement among the results produced by different methods (Poucet, 1989).

The need to model human errors, either quantitatively or qualitatively, resulted in a rather large number of HRA methods. Common to them all, however, was the focus on incorrect or faulty human actions. Associated with these methods were several models and conceptual frameworks, the most important being the skill-based, rule-based, knowledge-based (SRK) framework proposed by Rasmussen (1986). Although the SRK framework itself was not intended as a model of human error, it became the basis for one of the more important models of this kind; namely, the generic error modeling system (GEMS) (Reason, 1990). See Figure 3.2.

After TMI, the study of HRA became a necessary part of PRA, and the focus of HRA was for about three decades solely on the quantification of human error probability (HEP). The reason for this was that HRA was used in risk assessment, which is in itself probabilistic. The established practice of HRA served the following three purposes:

1. *Human error identification*, to identify which human errors could occur in a given situation or scenario. (This was the qualitative part of HRA.)
2. *Human error quantification*, to calculate how likely or probable the human errors would be, in order to be able to calculate the probabilities for various outcomes as part of a PRA. (This was the quantitative part of HRA.)

3. *Human error reduction*, to find ways to reduce the error likelihood by enhancing human reliability, if appropriate.

The fact that HRA had both a qualitative and a quantitative side was recognized by Swain (1990), who defined the qualitative part to be more or less what Kirwan (1994) called human error identification. This qualitative part can be used as the basis for design or redesign of a system. However, as Kirwan noted, the focus has been on human error quantification, and human error identification has received less attention and development resources than it rightly deserves.

Kirwan (1994) further pointed out that in the 1960s the HEP was calculated with the help of databases containing the failure rates for operators carrying out particular tasks. This way of solving the task was in the highest degree inspired by the way that reliability engineers estimated the risk of hardware failure: by simply looking at failure rates. Few of the databases with operator failure rates remain in use today, because it gradually has become obvious that humans differ significantly from hardware components in the sense that they neither work with very specific inputs and outputs nor are limited to one or a few functions. Technical systems and components are designed and built to perform with little or no variability, until they have to be replaced. Humans cannot naturally provide the same constant performance, nor should they rightly be expected to. It has in particular become clear that the human mind is not an information processing system.

First and Second Generation HRA

After the rapid development of HRA in the 1980s, partly to address the need created by the TMI accident, HRA reached a plateau where little substantial progress was made, despite a constant trickle of new methods. This led one of the leading practitioners of HRA to publish a paper in which he critically analyzed the established practices (Dougherty, 1990). This criticism pointed out that the established HRA methods suffered from a number of shortcomings, summarized by Swain (1990) as follows:

1. *Less-than-adequate data.* There is a scarcity of data on human performance that are useful for quantitative predictions of human behavior in complex systems.
2. *Less-than-adequate agreement in use of expert judgment methods.* There has not yet been a demonstration of satisfactory levels of between-expert consistency, much less of accuracy of predictions.
3. *Less-than-adequate calibration of simulator data.* Because the simulator is not the real world, the problem remains how raw data from training simulators can be modified to reflect real-world performance.
4. *Less-than-adequate proof of accuracy in HRAs.* Demonstrations of the accuracy of HRAs for real-world predictions are almost nonexistent. This goes particularly for nonroutine tasks (e.g., time-dependent diagnosis and misdiagnosis probabilities in hypothetical accident sequences).

5. *Less-than-adequate psychological realism in some HRA approaches.* Many HRA approaches are based on highly questionable assumptions about human behavior and/or the assumptions inherent in the models are not traceable.
6. *Less-than-adequate treatment of some important performance-shaping factors (PSFs).* Even in the better HRA approaches, PSFs such as managerial methods and attitudes, organizational factors, cultural differences, and irrational behavior are not adequately treated.

Dougherty's critical paper also introduced the terms "first generation" and "second generation" HRA, where the established methods (by 1990) by definition were the first generation. This challenge led to the development of several methods that became known as the second generation of HRA, of which the best known are ATHEANA (Cooper et al., 1996), CREAM (Hollnagel, 1998), and MERMOS (Bieder et al., 1998). The principal difference between the first and second generation methods may described as follows.

First generation methods focused on human error as the most important phenomenon and attempted to provide a quantitative estimate of that (i.e., the HEP). It was, however, always recognized that humans always acted in a work situation or context and that this might influence the HEP. The influence was expressed by means of the PSFs, which also were quantified and combined into a single value that was used to adjust the HEP.

The use of first generation HRA methods tacitly assumed that it was meaningful to consider each action by itself and to refer to error types, that it was possible to find the failure probability for error types, and that PSFs were mathematically independent, hence could be expressed by a single value. It is always necessary for a method to make some simplifying assumptions, but it is important that those assumptions have a reasonable degree of realism (Hollnagel, 2009). For first generation HRA, it was eventually acknowledged that the assumptions were either unrealistic or wrong, as summarized above. A more realistic set of assumptions would have to acknowledge that human actions are always part of a whole (a plan or a sequence), that actions always take place in a context, that there are no normal conditions for human actions because humans as individuals differ from one another, that there consequently is no inherent failure probability for individual actions, and finally that PSFs are not independent of each other.

Second generation HRA consequently changed the focus from the individual to the working conditions or context. A number of methods were developed to characterize the context and on the basis of that derive estimates of the probability that a human operator would fail to perform as intended and expected:

- In ATHEANA (A Technique for Human Error ANAlysis; Cooper et al., 1996), an unsafe act (slips, lapses, mistakes, and circumventions) was the specific human action that led to a human failure event. The unsafe act was triggered by error forcing contexts (EFCs), which expressed the combined effects of PSFs and plant conditions.
- In CREAM (cognitive reliability and error analysis method; Hollnagel, 1998), performance reliability was a function of the control mode, which in turn depended on the common performance conditions. The control mode

could have been scrambled, opportunistic, tactical, or strategic, depending on how much time was available to prepare and execute the action.

- In MERMOS (*méthode d'evaluation de la réalisations des missions opérateur pour la sûreté*; Bieder et al., 1998), performance reliability was determined by the CICAs (*configuration importante de la conduite accidentelle*), which corresponded to a functioning pattern of the operating system through time, consisting of the configuration (organization mode) and the orientation (positioning with respect to the situation). (A more detailed description of MERMOS is provided in Chapter 4.)

Another way of explaining the difference between first and second generation HRA methods is to consider the distinction between signal and noise, as described by, for example, the theory of signal detection (e.g., Swets, 1964). In first generation HRA, the HEP was the signal and the PSFs were the noise (Figure 3.3). In other words, the methods focused on individual human actions and the efforts were directed at finding the data to support the theories and models. This meant data about human failures and human error rates, much as one could find reliability data for technology. Unfortunately, the assumption that a function can simply be repeated until it fails does not work for humans, because the underlying premise that humans are machines and function in a uniform manner is fundamentally wrong (see theory W later in this chapter). Examples of this approach are found in conventional human information processing and in the error psychology that looks at error types rather than error modes.

In second generation HRA, the roles changed in the sense that the performance conditions or the context became the signal and the individual human failure or error became the noise (see Figure 3.4). In other words, it was acknowledged that

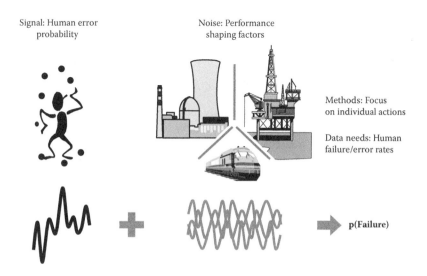

FIGURE 3.3 Signal and noise in first generation on HRA.

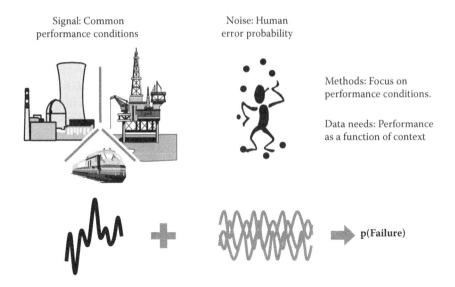

FIGURE 3.4 Signal and noise in second generation HRA.

the variability and effect of the working conditions in most cases dominated the variability of individual human performance. Or, to put it even more strongly, it was realized that there were no human errors as such, but that people always acted as well as they could given the circumstances. This led to a change in the methods, which now focused on the performance conditions. The data needs also changed and became how performance could be described vis-à-vis, and as depending on, the context. Examples of this approach are found in the MTO (man, technology, organization) framework, in the thinking about organizational accidents, in cognitive systems engineering and joint cognitive systems, and even in some classical HRA methods such as the TRC (where the probability of responding is a function of time) and HEART (which refers to task conditions rather than tasks.)

THINKING ABOUT HOW THINGS CAN GO WRONG

In its second role, human factors was mainly motivated by a need to understand how things could go wrong and how this could be avoided or prevented. Most efforts were therefore spent on understanding how accidents happened. This was in most cases synonymous to developing theories or models about possible causes, as well as proposing methods for finding the causes.

SEQUENTIAL ACCIDENT MODELS

A classic sequential model is the domino theory described by Heinrich, Petersen, & Roos (1980). The domino model describes an accident as a chain of causes and effects, where an unexpected event may initiate a sequence of consequences of which the last ones are the accident and the injury (see Figure 3.5). According to

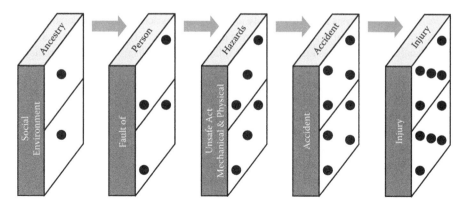

FIGURE 3.5 The domino model (after Heinrich et al., 1980).

the domino theory, an accident can be avoided if one or more domino blocks are prevented from falling. The domino model also implies that it is possible to reason backwards from the final outcome, the injury, to the initiating event, hence the original or root cause.

The role of determinism in accidents has probably never been expressed more clearly than in the axioms of industrial safety that provide the justification for the domino model. The first of these axioms reads as follows:

> The occurrence of an injury invariably results from a completed sequence of factors — the last one of these being the accident itself. The accident in turn is invariably caused or permitted directly by the unsafe act of a person and/or a mechanical or physical hazard. (Heinrich, Petersen, & Roos, 1980, p. 21)

The first axiom describes a strictly deterministic situation, in which the cause always is the unsafe act of a person or a mechanical or physical failure. According to this line of thinking it therefore makes excellent sense to look for the cause.

ROOT CAUSE DEFINED

The search for the causes of accidents can be based on either a single cause philosophy, as in the domino mode, or a multiple cause philosophy. A single cause philosophy corresponds to the straightforward interpretation of root cause analysis (i.e., that there is a single cause for any outcome that, if acted upon, will prevent the outcome from occurring). The root cause thus dominates all other contributing factors. Although the single cause philosophy is tempting, not least because it provides for an effective analysis, it is clearly also a potentially dangerous oversimplification, because it may lead to incorrect solutions. The alternative is a multiple cause philosophy, which accepts that an outcome may be the result of a combination of a number of factors and that a root cause can exist for each of these contributing factors. In common with the single cause philosophy, it is assumed the undesired outcome can be prevented by acting on the necessary causes. In other words, the basic assumption of causality — or cause–effect relations — is the same in both philosophies.

A root cause analysis is more formally any structured process used to understand the causes of past events for the purpose of preventing recurrence. To be classified as a root cause analysis, the process must answer the following questions in the order presented:

1. What is the problem to be analyzed? (problem definition)
2. What is the significance to the stakeholders? (significance)
3. What are the causal relationships that combined to cause the defined problem? (problem analysis)
4. How are the causes interrelated? (causal chart)
5. What evidence is provided to support each cause? (cause validation)
6. What are the proposed solutions to prevent recurrence? (corrective actions)
7. What assurance is provided that the solutions will prevent recurrence? (solution validation)

EPIDEMIOLOGICAL ACCIDENT MODELS

Epidemiological accident models are based on an analogy with what happens when a person is exposed to an infection. The host–agent–environment model (Figure 3.6) is used to describe the processes during infections. An agent (e.g., a virus) is able to successfully infect a host if the defense (i.e., the immune system) is defective. This analogy is used by the epidemiological accident model to describe how latent conditions (weaknesses of the defense system) in combination with active failures (the agent) can adversely affect the system (the host), thus creating an accident. Epidemiological accident models are clearly still sequential accident models. But instead of having only active failures (events), the model combines active failures and latent conditions. The latent conditions may in turn be the effect of a failure that happened long before the now active failure that seemingly triggered the accident.

Epidemiological accident models consider accident development as linear but extend the scope in terms of time and levels of analysis (e.g., latent failures that originate from design can be included in the account). Epidemiological accident models

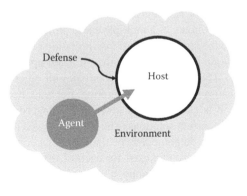

FIGURE 3.6 The epidemiological accident model.

are complex linear models and differ from the simple linear models, such as the domino model, in four important ways:

1. They use the more neutral term "performance deviation" instead of terms such as "unsafe acts" which have come to be treated as synonymous to human error. This shifts the focus away from humans, because performance deviations could be applied to individuals and social systems as well as to technical components.
2. The environment is included in the analysis and is assumed to affect humans, social systems, and technical components. This enlargement makes the model more open-ended than the sequential models and their search for a single root cause.
3. Barriers are introduced as a way to prevent or stop an accident from evolving. This offers an alternative to the elimination of causes, which is the main solution in simple, linear models.
4. Latent conditions (at first called latent failures) are recognized as important components of accidents.

The best known example of an epidemiological accident model is unquestionably James Reason's famous Swiss cheese analogy (Reason, 1997). In this model, the cheese slices illustrate layers of defense, barriers, and safeguards. The holes in the slices of cheese illustrate defense failures or weaknesses. If and when the holes in the barriers are aligned, an accident can occur. (A more detailed analysis of this model can be found in Reason, Hollnagel, & Paries, 2006.)

THE HUMAN ERROR BIAS

In accident analysis the law of causality reigns supreme, but mostly in the shape of its logical opposite, the law of reverse causality. Whereas the law of causality states that every cause has an effect, the reverse law states that every effect has a cause. Although this is philosophically reasonable, in the sense that it is unacceptable that events happen by themselves, it is not a defensible position in practice. Even worse, assuming that a cause must exist does not guarantee that it can be found. The belief that this is possible rests on two assumptions: the law of reverse causality and the rationality assumption (the idea that it is logically possible to reason backwards in time from the effect to the cause). Quite apart from the fact that humans are notoriously prone to reason in ways that conflict with the rules of logic, the rationality assumption also requires a deterministic world that does not really exist.

This first axiom of industrial safety, which obviously was no more than an expression of the accepted wisdom among practitioners at the time, expresses two important assumptions; namely, those of reverse causation and that human errors are the root causes of accidents. The belief in the existence of a cause or causes and in our ability to find them, nevertheless does not mean that human error is a necessary construct. Before the mid-1950s, for instance, people were quite satisfied if and when a technical cause could be found (i.e., that something had broken or did not work). Indeed, much of technology was previously unreliable, at least relative to our present standards, and it was therefore perfectly reasonable to look for the cause there. This changed drastically around the 1950s when the availability of affordable digital technology made it possible for many processes to be controlled without relying on humans. As a result of that, the speed by which things could be done increased significantly and continues to increase to this day. The effects of computerization, added to those of mechanization, centralization, and automation (Hollnagel & Woods, 2005), however, meant that humans were faced with a new kind of work for which they were ill suited. In a paradoxical way, technology was again limited by human performance, although not on the levels of psycho-motor functions, but on the levels of monitoring and planning.

Human factors engineering came into being as a way of solving these problems, and led to a broad acceptance of the view that the human operator was a major weak link in the control of processes and in socio-technical systems in general. Accident investigators often seemed to assume that the process as such was infallible, or would have been so had the operators not done something wrong. It was therefore natural to stop the analyses once a human error had been found. This was elegantly expressed by Perrow (1984), who wrote:

> Formal accident investigations usually start with an assumption that the operator must have failed, and if this attribution can be made, that is the end of serious inquiry. Finding that faulty designs were responsible would entail enormous shutdown and retrofitting costs; finding that management was responsible would threaten those in charge, but finding that operators were responsible preserves the system, with some soporific injunctions about better training. (p. 146)

As this quote demonstrates, the human error bias is singularly nonconstructive for accident analysis.

TRACTABILITY AND INTRACTABILITY

The established safety analysis methods, including methods for HRA, share a number of assumptions. These assumptions are a heritage from the large-scale technological systems and conditions for which the first safety assessment methods were developed in the late 1950s, such as the Minuteman missile system. Although the underlying assumptions rarely are stated explicitly, they are easy to recognize by taking a second look at established methods and tools, such as FMEA, HAZOP, fault trees, etc. (see Figure 3.7). The four main assumptions are:

1. *A system can be decomposed into meaningful elements* (typically components or events). Although this clearly is warranted for technological systems, the assumption is not necessarily valid for social systems and organizations, or for human activities (tasks). The principle of decomposition is in conflict with the principle that the whole is larger than the sum of the parts.
2. *The failure probability of elements can be analyzed and described individually.* This is the rationale for focusing on the human error probability,

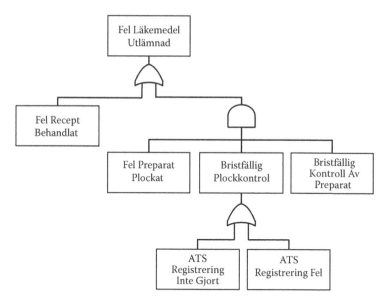

FIGURE 3.7 A fault tree.

and indeed for classifications of human errors. (It is also the rationale for root cause analysis.) Although this assumption may be valid for technological systems, it is definitely not so for humans and organizations — even if meaningful components could be found.

3. *The order or sequence of events is predetermined and fixed.* If a different sequence of events needs to be considered, it is necessary to draw a different diagram or tree. This assumption is clearly illustrated by the fault tree in Figure 3.7.

4. *Combinations can be described as linear* (tractable, noninteracting). As illustrated by Figure 3.7, combinations are either disjunctions (OR gates) or conjunctions (AND gates).

The last assumption introduced the two terms: *tractable* and *intractable*. The terms are used to characterize systems and organizations with regard to whether it is possible to know what goes on inside them (i.e., whether a sufficiently clear description or specification of the system and its functions can be provided). This requirement must, for instance, be met in order for a system to be analyzed, in order for its risks to be assessed, and for its safety to be managed. That this must be so is obvious if we consider the opposite. If we do not have a clear description or specification of a system, and/or if we do not know what goes on inside it, then it is impossible effectively to control it or to make a risk assessment. These qualities are captured by making a distinction between tractable and intractable systems (see Table 3.1).

The established human factors methods all require that it is possible to describe the human–machine system in detail; for instance, by referring to a set of scenarios and a corresponding required functionality. In other words, the human–machine

TABLE 3.1

Tractable and Intractable Systems

	Tractable System	Intractable System
Number of details	Descriptions are simple with few details	Descriptions are elaborate with many details
Comprehensibility	Principles of functioning are known	Principles of functioning are partly unknown
Stability	System does not change while being described	System changes before description is completed
Relationship to other systems	Independence	Interdependence
Metaphor	Clockwork	Teamwork

system must be tractable. Most socio-technical systems, including nuclear power plants, are unfortunately intractable. This means that methods and approaches that assume that the human–machine system is tractable will be unsuitable for accident analysis and safety management. It is therefore necessary to look for methods and approaches that can be used for intractable systems (i.e., for systems that are incompletely described or underspecified).

THEORY W

The current approaches to human–machine system safety echo the traditional human factors perspective that humans are imprecise, variable, and slow, and that furthermore human performance variability is a leading cause of accidents. Another way of expressing that is by pointing out that both accident analysis and risk assessment seem to subscribe to a particular view of safety, which we can call theory W. According to theory W, systems work because the following conditions are met:

- Systems are well designed and scrupulously maintained.
- The procedures that are provided are complete and correct.
- People behave as they are expected to, and more important, as they have been taught or trained to do.
- System designers have been able to foresee and anticipate every contingency.

Theory W describes well-tested and well-behaved systems, or in other words, tractable systems. Characteristic of such systems is a high degree of equipment reliability; workers and managers who are vigilant in their testing, observations, procedures, training, and operations; well-trained staff; enlightened management; and good operating procedures in place. Under such assumptions, humans are clearly a liability and their inability to perform in a reliable (or machine-like) manner is a threat. According to theory W, safety can be achieved by constraining all kinds of performance variability so that efficiency can be maintained and malfunctions or

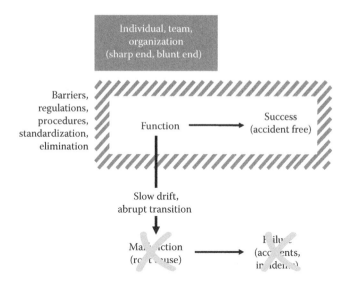

FIGURE 3.8   Safety according to theory W.

failures avoided (Figure 3.8). Examples of frequently used constraints are strict training, barriers of various kinds, procedures, standardization, rules, and regulations.

The notions of success and failure in theory W are, of course, an oversimplification. They represent the general idea that some outcomes are deemed acceptable whereas others are deemed unacceptable. The criteria for acceptability and unacceptability are not always crisp, but may vary with the conditions — either local (acute or short term) or global (permanent or longer term). Yet, under any given circumstances there will definitely be some outcomes that are clearly acceptable and some that are clearly unacceptable — possibly with a gray zone between them. For the sake of this discussion we shall assume that it is meaningful to talk about successes and failures.

THE HUMAN FACTOR AS AN ASSET

Although theory W for some may represent an ideal, it is not achievable in practice. The two main reasons are, first, that all systems of interest are more or less intractable, and second, that performance variability is inevitable.

Nuclear power plants and air traffic management, as well as many other present-day systems of major interest for industrial safety, are intractable. For such systems, the principles of functioning are only partly known, descriptions are elaborate and contain many details, descriptions take a long time to make, and the system therefore changes before the description can be completed. Consequently, it is never possible to provide a complete description or specification of the system.

Intractable systems are underspecified in the sense that details may be missing or unavailable (e.g., Clarke, 2000). If a system is underspecified it is clearly not possible to provide precise procedures or instructions. On the contrary, the people working in the system, be it at the sharp end or at the blunt end, must be able to use the available prescriptions and procedures in actual situations that differ from what was assumed. In other words, it is necessary that people are able to vary or adjust what they do to ensure that the system functions as required and achieves its operational goals. Performance variability — whether it is called improvisation, adaptation, adjustments, efficiency–thoroughness trade-off, sacrificing decisions, or creativity — is therefore an asset rather than a threat, and methods for risk analysis and safety management must be able to take that quality into account. The established approaches are, however, on the whole, incapable of achieving that.

PERFORMANCE VARIABILITY

Although machines and technological artifacts are designed, built, and maintained so that they can produce a near constant performance, at least until they fail and must be replaced, the same is not the case for humans and for social systems (organizations). Human performance is always variable, for a number of reasons.

As already mentioned, performance variability, in the form of habitual and/or intentional adjustments of performance, is necessary because performance conditions as a rule are underspecified. This leads to a nearly unavoidable difference between work-as-imagined and work-as-actually-done (see next section on theory Z). Performance variability is, however, on the whole, more often a strength than a liability, and often it is the primary reason why socio-technical systems work as well as they do. Humans are extremely adept at finding effective ways of overcoming problems at work, and this capability is crucial for both safety and productivity. Human performance can therefore at the same time both enhance and detract from system safety. Assessment methods must be able to address this duality.

In addition to the variability coming from intentional or habitual performance adjustments, performance variability is also produced by a number of internal and external factors. The six main sources of human performance variability are:

1. Inherent physiological and/or fundamental psychological characteristics. Examples are fatigue, circadian rhythm, vigilance and attention, refractory periods, forgetting, associations, etc.
2. Higher level psychological phenomena such as ingenuity, creativity, and adaptability; for instance, in overcoming temporal constraints and underspecification.
3. Organizational factors, such as external demands (quality, quantity), resource stretching, goal substitution, etc.
4. Social factors, such as the expectations of oneself or of colleagues, compliance with group working standards, etc.
5. Contextual factors, such as working conditions that are too hot, too noisy, too humid, etc.
6. Unpredictability of the domain; for example, weather conditions, number of flights, pilot variability, technical problems, etc.

Theory Z

Altogether, this means that performance variability is inevitable, both on the level of the individual and on the level of the social group and organization. At the same time, performance variability is also needed, as argued in the previous section. In recognition of this, in the third role the human factor is an asset rather than a liability. This role has, however, not come about abruptly, but has been slowly growing for the last fifteen years or so. To be consistent with this role, accident analysis and risk assessment should not be based on theory W, but rather on something that we can call theory Z. According to theory Z, systems work because the following conditions are met:

- People learn to identify and overcome design flaws and functional glitches.
- People can recognize the actual demands and adapt their performance accordingly.
- When procedures must be applied, people can interpret and apply them to match the conditions.
- Finally, people can detect and correct when something goes wrong or when it is about to go wrong, and hence intervene before the situation seriously worsens.

Theory Z describes work as actually done rather than work as imagined; hence systems that are real rather than ideal. Such systems may still be very reliable, but this is because people are flexible and adaptive, rather than because the systems have been perfectly thought out and designed. Under such assumptions, humans are no longer a liability and performance variability is not a threat. On the contrary, the variability of normal performance is necessary for the system to function, and it is the source of successes as well as of failures. (In theory Z, failures are also seen as opportunities for learning and thus as having a positive as well as a negative value.) Because successes and failures both depend on performance variability, failures cannot be prevented by eliminating this performance variability; in other words, safety cannot be managed by imposing constraints on normal work. The solution is instead to identify the situations where the variability of normal performance may combine to create unwanted effects and continuously monitor how the system functions in order to intervene and dampen performance variability when it threatens to get out of control (see Figure 3.9). (Conversely, performance variability should be accentuated or amplified when it can improve successful outcomes.) This leads to the principle of safety by management, which is at the heart of resilience engineering.

RESILIENCE ENGINEERING

Because performance variability is both normal and necessary, safety must be achieved by controlling performance variability rather than by constraining it. This principle is encapsulated in resilience engineering (Hollnagel, Woods, & Leveson, 2006). According to this, a resilient system is defined by its ability to adjust its functioning prior to or following changes and disturbances so that it can go on working

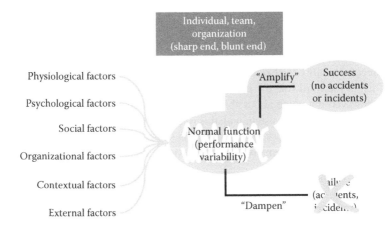

FIGURE 3.9 Safety according to theory Z.

even after a major mishap or in the presence of continuous stress. The quality of resilience can be defined more precisely by pointing to four essential qualities or abilities that a system or an organization must have (see Figure 3.10):

1. *Respond to regular and irregular threats in a robust, yet flexible, manner.* It is not enough to have a ready-made set of responses at hand, because actual situations often do not match the expected situations — the only possible exceptions being routine normal operation. The organization must be able to apply the prepared response such that it matches the current conditions both in terms of needs and in terms of resources. In terms of the three types of threats proposed by Westrum (2006), this is the ability to deal with regular threats. The response enables the organization to cope with the *actual*.

2. *Flexibly monitor what is going on, including its own performance.* The flexibility means that the basis for monitoring must be assessed from time to time, to avoid being trapped by routine and habits. The monitoring enables the organization to cope with that which is, or could become, *critical* in the near term.

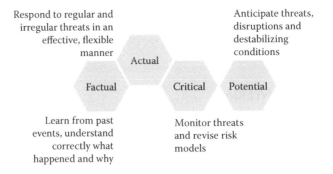

FIGURE 3.10 The four main principles of resilience engineering.

3. *Anticipate disruptions, pressures, and their consequences that lie further ahead.* This means the ability to look beyond the current situation and the near future, and to consider what may happen in the medium to long term. In terms of the three types of threats proposed by Westrum, this is the ability to deal with irregular threats, possibly even the unexampled events. The anticipation enables the organization to cope with the *potential*.

4. *Learn from experience.* This sounds rather straightforward, but a concrete solution requires consideration of which data to learn from, when to learn, and how the learning should show itself in the organization — as changes to procedures, changes to roles and functions, or changes to the organization itself. The learning enables the organization to cope with the *factual*.

INDIVIDUALS AND ORGANIZATIONS

In resilience engineering (Hollnagel, Woods, & Leveson, 2006), resilience is a quality of the system or the organization, rather than of the individual. An organization can be defined as the planned, coordinated, and purposeful action of human beings to construct or compile a common tangible or intangible product (i.e., a social arrangement that pursues collective goals, controls its own performance, and has a boundary separating it from its environment). (The word *organization* is derived from the Greek word *organon,* meaning tool.) An organization is a certain structure or arrangement of people. An organization can therefore be resilient by virtue of the people in it or by virtue of the social arrangement.

Although the resilience in a very fundamental sense depends on the people and on the variability of their normal, it is clearly impractical, if not impossible, for a single individual to provide the four essential abilities that are the mark of a resilient system (see Figure 3.10). This can, however, be achieved by the organization, by the social arrangement of work. An organization can assign different functions to individuals and groups, provide the resources necessary to implement the functions, monitor the progress, and integrate and apply the outcomes. The organization is therefore needed to provide the resilience that is necessary not only for its own existence but also for its ability to produce what is expected of it under a wide range of conditions.

Resilience engineering clearly recognizes — and promotes — the third role of the human factor, as a source of safety rather than as a threat. This role must constantly be reinforced on the individual level; for instance, by becoming part of the organizational culture. It must be recognized in everything the organization does, not the least in the way it looks at adverse events (accidents, incidents, etc.). The organization can become resilient only if the social arrangement of work enables effective coping with the actual, the critical, the potential, and the factual, and if the culture and norms encourage the view of the human factor as a source of safety and strength. The third role of the human factor means not only that the perspective has changed from a focus on what can go wrong to a focus on what can go right. It also means that the traditional notion of an exclusive human factor, as something that can be addressed by itself, is replaced by a comprehensive view of the socio-technical system as a whole. To be successful, safety in complex industrial environments must be developed on that level, and the new human factors must help by developing the tools and theories to make that possible.

REFERENCES

Bieder, C., Le-Bot, P., Desmares, P. & Bonnet, J.-L. (1998). MERMOS: EDF´s new advanced HRA method. *Proceedings of PSAM-4*. London: Springer-Verlag.

Clarke, S. G. (2000). Safety culture: Underspecified and overrated? *International Journal of Management Reviews*, 2(1), 65–90.

Cooper, S. E., Ramey-Smith, A. M., Wreathall, J., Parry, G. W., Bley, D. C. & Luckas, W. J. (1996). *A technique for human error analysis (ATHEANA)*. Washington, DC: U.S. Nuclear Regulatory Commission.

Dougherty, E. M. Jr. (1990). Human reliability analysis — Where shouldst thou turn? *Reliability Engineering and System Safety*, 29(3), 283–299.

Fitts, P. M. (1951). *Human engineering for an effective air navigation and traffic control system*. Ohio State University Foundation Report, Columbus, OH.

Hale, A. R. (1978). *The role of HM inspectors of factories with particular reference to their training*. PhD thesis. University of Aston in Birmingham.

Heinrich, H. W., Petersen, D. & Roos, N. (1980). *Industrial Accident Prevention*. New York: McGraw-Hill.

Hollnagel, E. (1998). *Cognitive reliability and error analysis method*. Oxford: Elsevier Science.

Hollnagel, E. (2000). Looking for errors of omission and commission or the hunting of the Snark revisited. *Reliability Engineering and System Safety*, 68(2), 135–145.

Hollnagel, E. (2009). *The ETTO principle: Why things that go right sometimes go wrong*. Aldershot, UK: Ashgate.

Hollnagel, E. & Woods, D. D. (2005). *Joint cognitive systems: Foundations of cognitive systems engineering*. Boca Raton, FL: Taylor & Francis.

Hollnagel, E., Woods, D. D. & Leveson, N. (Eds.) (2006). *Resilience engineering: Concepts and precepts*. Aldershot, UK: Ashgate.

Jastrzebowski, W. B. (1857). An outline of ergonomics or the science of work based on the truths drawn from the science of nature. *Przyroda i Przemysl (Nature and Industry)*, 2, 29–32.

Kirwan, B. (1994). *A guide to practical human reliability assessment*. London: Taylor & Francis.

Leveson, N. G. (1994). High pressure steam engines and computer software. *IEEE Computer*, October.

Marmaras, N., Poulakakis, G. & Papakostopoulos, V. (1999). Ergonomic design in ancient Greece. *Applied Ergonomics*, 30(4), 361–368.

Miller, R. B. (1953). *A method for man–machine task analysis*. Dayton, OH: Wright AFB Development Center.

Perrow, C. (1984). *Normal accidents: Living with high risk technologies*. New York: Basic Books.

Poucet, A. (1989). *Human factors reliability benchmark exercise — Synthesis report*. Ispra, VA: CEC Joint Research Centre.

Rasmussen, J. (1986). *Information processing and human–machine interaction*. New York: North-Holland.

Reason, J. T. (1990). *Human error*. Cambridge: Cambridge University Press.

Reason, J. T. (1997). *Managing the risks of organizational accidents*. Aldershot, UK: Ashgate.

Reason, J. T., Hollnagel, E. & Paries, J. (2006). *Revisiting the "swiss cheese" model of accidents*. Paris: Eurocontrol Experimental Centre.

Swain, A. D. (1990). Human reliability analysis: Need, status, trends and limitations. *Reliability Engineering and System Safety, 29*, 301–313.

Swets, J. A. (Ed.) (1964). *Signal detection and recognition by human observers.* New York: Wiley.

Taylor, F. W. (1911). *The principles of scientific management.* New York: Harper.

Westrum, R. (2006). A typology of resilience situations. In E. Hollnagel, D. D. Woods & N. Leveson (Eds.). *Resilience engineering: Concepts and precepts.* Aldershot, UK: Ashgate.

Wiener, N. (1993). *Invention: The care and feeding of ideas.* Cambridge, MA: MIT Press.

4 The Meaning of Human Error in the Safe Regulation Model for Risky Organizations

Pierre Le Bot

CONTENTS

INTRODUCTION

Having to extend the area of application of our probabilistic human reliability assessment (HRA) methods to organizational situations other than the accidental operation of a reactor or to other at-risk organizations, we have had to consider the impact of organizational factors. The design of the safe regulation model has helped us to give a response to these two problems. This model is based on the notion of effective rules according to Reynaud's theory of social regulation (Reynaud, 1989) and aims at specifying the notion of effective rules in the field of risky, yet ultra-safe industries such as nuclear power generation. One first important result from the safe regulation model is a clarification of the concepts of human error and its role in accidents, as it is implemented in our human reliability method called *méthode d'evaluation de la réalisations des missions opérateur pour la sûreté,* or MERMOS (Le Bot et al., 1999).

THE THEORY OF SOCIAL REGULATION

The safe regulation model is based on the theory of social regulation (in French, *théorie de la régulation sociale,* or TRS), from Reynaud (1989). This model, which we are currently building within the human reliability team of Électricité de France's (EDF) research and development department, explains the impact of organizational factors on the safety of the operation of ultra-safe systems at risks such as nuclear power plants. TRS bases the understanding of social relations (particularly in the working environment) on social regulations. Social regulation is the production of social rules governing the behavior of collectives and the updating of these rules. It has been applied to the modeling of operating teams in normal operation at nuclear plants by de Terssac (1992). This theory emphasizes several social regulations that produce collective rules.

Control regulation (*régulation de contrôle* in French), which is representative of the management of work most of the time, provides the preexisting control rules prior to the action; for example, the procedures and principles of organization of the action. We can define two kinds of events where a safe action is needed. The first kind of event is related to planned situations where an intervention is executed to maintain or operate the system. The second kind of event is related to unplanned situations for which emergency action is needed to mitigate the consequences of a degradation of the state of the system. Autonomous regulation (*régulation autonome* in French) in both cases operates in the situation during the action. The actors use their know-how and knowledge to establish the technical rules appropriate to the situation. But as Reynaud (1989) has shown, a third collective procedure comes into play to arrive at a compromise between the two types of rules. A negotiation procedure between actors representing each type of regulation results in the setting up of effective rules, between the control rules and the autonomous rules, in order to satisfy the requirements of both as well a possible. For Reynaud, this joint regulation is exercised *after* the event, *after* the action, following the accomplishments, according to de Terssac (1992). In the safe regulation model we will modify that definition: an important issue for resilient systems is to be able to jointly adapt control rules and autonomous rules both *in* a situation (in real time), on the one hand, and *after*

TABLE 4.1

The Different Types of Regulation

Regulation	When Does it Work?	Rules Produced	Instances of Regulation	Transmission of the Rules (or the Capacity of Designing a Rule)
Control regulation	Before the situation	Doctrine and prescriptions of the operation Organization of the roles and functions Delegation in situation control	Legal regulations Design Management	Procedures Training Interface Safety culture Hierarchy
Autonomous regulation	In a situation (or just before the situation)	Know-how Practices	Work collectives	On-the-job learning Experiences Collectives History
Joint regulation in a situation	In a situation or just before the situation during the rupture phase (during reconfiguration in MERMOS)	Effective rules (CICAs in MERMOS) It should be noted that an effective rule may simply be a control or autonomous rule, selected and validated	Work collectives	The effective rules are destroyed or adapted by the deferred joint regulation after validation
Deferred joint regulation	After the situation	Feedback for the situations experienced translates into new control rules or autonomous rules	Feedback process integrating control and autonomous regulation	Capitalization of feedback

the situation has occurred (deferred in time), on the other hand. Table 4.1 summarizes the three regulations that will be detailed here, the joint regulation of Reynaud being decomposed for the safe regulation model to in-situation joint regulation and deferred joint regulation.

CONTROL REGULATION

With control regulation the organization anticipates the risk. By designing validated rules before an event, whether planned or unplanned, the management provides the operators with operative organization, knowledge, and optimized objectives to control the situation. (In the case of unforeseen or probabilistic events, the validated rules are the emergency operating procedures.) Operators cannot and should not design complex rules during an event, as it develops. They cannot and should not act only from their experience, and improvisation could be dangerous: they need the organizational and technical knowledge provided by the control rules. Therefore, before actors are confronted with the events, the organization has to elaborate the relevant control rules by taking into account the operational or simulated feedback and theoretical calculations. If this is done, then the actors in the situation at the sharp end are not alone; they interact with their procedures, their training and safety culture, and their interface. We have to take into account that the normal or emergency operation emerges from that interaction. The TRS considers a group of social actors as the object and the source of regulation: for the safe regulation model, we consider the group of actors concerned directly in a situation by an operation, interacting with its procedures and its interface. We call that system the operation system. For an easier understanding we can also use the expression "the actors" with the same meaning.

In order to build the safe regulation model, we can specify how the control rules have to define the task devolved to the operation system and the way to achieve it. We can state that the control rules have to cover the following six dimensions exhaustively: diagnosis, objectives, prioritization, selection of means, configuration, and surveillance.

DIAGNOSIS

The rules of diagnosis allow the actors to recognize the type of event they deal with. The types of event are predefined depending on the physical parameter levels and materials availability (state diagnosis), and on the causes, consequences, evolution of the event, and urgency to act (situation diagnosis).

OBJECTIVES

The rules of objectives set the goals that the actors must reach in order to:

1. Mitigate the potential consequences of the situation if it is degraded (mitigation)
2. Prevent a further degradation from a new aggravating event (prevention)
3. If necessary, improve the situation if it is degraded, by eliminating the causes of degradation or canceling their effects (improvement)
4. Optimize task execution in the most efficient way by saving resources and time (efficiency)

PRIORITIZATION

The prioritization rules serve to set the priority among the rules of objectives, given the diagnosis. Objectives can require incompatible configurations of the system (for example, temperature stabilization or cool down). Moreover, the resources of the actors (manpower, time, communication tools, etc.) and the resources of the system (means of control of the thermo-hydraulic state, energy, etc.) are limited. The rules of priority then determine which objective has the priority, in order to decide how to configure the system and organize the actors and their tasks (see the rules in the configuration section).

For an accident, the order of priority is generally from the most important to the least important: mitigation, prevention, improvement, efficiency. But if the situation is not degraded (during maintenance, for example), the objectives of mitigation and improvement are not relevant. Then the order of priority is reduced to prevention and efficiency.

SELECTION OF MEANS

When several means are available to perform the task, the rules of selection of means determine which means have to be chosen in order to comply with the former rules (for example, the urgency to act requires a quick response rather than an efficient one).

CONFIGURATION

The rules of configuration indicate how to organize the material and human resources in order to comply with the previous rules (diagnosis, objectives, prioritization, selection of means). One important aspect is the delegation in the situation of responsibility for management. This means that the organization explicitly has to hold somebody responsible for validation, in a situation of the rules that will lead to decisions and actions. Very often that role is assigned to the chief of the team.

SURVEILLANCE

The rules of surveillance give the criteria that make it possible to detect, from a set of situation parameters and information from the system, if any of the rules becomes irrelevant. A rule is irrelevant if it is no longer required (for example, if an objective has been fulfilled) or if it is not valid for the situation (the situation has evolved and the rule is erroneous). If a criterion is met, the rules of surveillance should stop the actual rules and trigger a reconfiguration in order to adopt new valid rules.

IMPLEMENTING CONTROL REGULATION

How are the control rules implemented? A prescription written as a paper procedure is only a potential rule: words and drawings become rules in the situation with the interaction between actors and procedures. Initial knowledge and choices of the organization described on the paper are implemented in decisions and actions after the actors have selected this piece of paper and translated the symbols using their training and their safety culture. Similar interactions exist with the interface: when actors take into account an alarm to implement an action relative to the meaning of that alarm, the control rules emerge from this interaction and drive the behavior of the human collective interacting with procedures and interface. We can therefore consider four supports for the control rules: prescriptions, training, safety culture, and interface design.

PRESCRIPTIONS (PROCEDURES)

The prescriptions define the tasks and the roles of the actors and can be written on paper or be available as computerized procedures that are used during action. They can also simply be known by the actors without the need of material to implement them.

TRAINING

Through training, the actors learn the meaning of the prescriptions and how to implement them in the situations they may face. They can also directly learn prescriptions in such a way that they do not need a procedure to implement them in the situation.

SAFETY CULTURE

What is interesting for the safe regulation model in the fuzzy concept of safety culture is the ability of the actors to produce control rules or partial control rules in the situation. This ability concerns the understanding by the actors of the underlying goals of the organization in the prescriptions, in order to criticize and to replace these prescriptions. This ability is required when the prescription is missing or false (for example, a typing error in a procedure).

INTERFACE DESIGN

The way that an indicator is graduated, the threshold value of an alarm, or the categorization of an alarm are supports of control rules as soon as the actors detect, validate, and use them to act. In the same way, all the devices that prevent, inhibit, or delay an action are supports of control rules implemented by the actors, provided they respect their functioning.

The control rules are designed before the situation occurs. Then they are static and necessarily simplify the reality. To be accessible by training and procedures, they must not be too detailed and they cannot be optimized for every situation, even if they can provide the operators with an answer for most situations. They cannot take into account the management of time, because the complexity of the systems does not allow planned management of situations either that are unplanned or where a small change in a parameter can significantly affect the development of the situation. Moreover, control rules can be wrong for a particular situation because it is practically and economically impossible to describe every possible situation in detail. Control rules can be wrong because something is missing or because of a design error. An organization that would limit its means of regulation to the control prior to the situation could turn out to be too rigid and be unable to adapt itself to a lack of anticipation.

In order for an organization to be safe, it must be able to complete the control rules designed prior to the situation, and it must be able to combine the anticipations of the situations with the reactiveness and flexibility that come from the autonomy of the actors. The second regulation is autonomous regulation.

AUTONOMOUS REGULATION

Autonomous regulation concerns actors that are involved with the system at the sharp end and refers to the production by the actors of autonomous rules from experience and know-how. Autonomous regulation and the autonomous rules are usually known as work practices. Nathanael & Marmaras said (2008):

> Practices are more like a constellation of alternative ways of doing, which are profoundly embedded in their physical settings, and in the minds and bodies of the people that enact them. As such they resist detailed description even by practitioners themselves. Their existence is manifested mostly in action and evidenced, for example, by the effortless and successful changes in courses of action to cope with differing situations. In other words, the core invariant in repetitions of practices is not the recurrence of the same events as seen by an external observer, rather, it is a kind of convergent generative method of who, when and how to act which seems to be assimilated by a specific community in a specific setting.

The authors point out the main characteristics of work practices or, in our words, of autonomous regulation and autonomous rules:

1. The regulation is activated in the work situation (or just before), in contrast to control regulation. The autonomy of the actors allows them to elaborate (explicitly or implicitly), during or just before action, rules that are to be followed.
2. Actors use their individual and collective experience to determine what has to be done specifically in the context where they are, from similar situations that they or their colleagues have experienced. The interactions between the individual and the collective build the individual and collective competencies, with the professional skills and know-how as a common ground.
3. The competency is the ability to produce relevant rules fitted to the situation (and only to that situation). It is different from knowledge and safety culture, which are gained mainly through training and reflect management rationality.

We postulate, in the safe regulation model, that cooperation in a situation is necessary. Autonomous regulation contributes to define the effective rules via a negotiation with control regulation.

MERMOS AND CICAS

Before going on to describe the third kind of rules, the effective rules, a brief description of MERMOS and its main concept, the CICAs (*configuration importante de la conduite accidentelle,* or key configuration for accident management), is needed to understand the safe regulation model, because we have built it from the extension of the MERMOS concepts. EDF currently uses MERMOS as its reference method for analyzing and quantifying the probabilities of a failed emergency operation in the control room after an incident or an accident has been triggered (Le Bot et al., 1999). MERMOS takes into account the whole of the emergency operation at the level of operators, procedures, and human–machine interface (i.e., the operation system, as for the safe regulation model). Individual operator errors are considered as part of a particular context that promotes a consistent behavior by the system. Sometimes this behavior would be appropriate if the situation was slightly different, despite being inappropriate for the triggered situation. The objective of MERMOS is thus to analyze and quantify failures in the reactor emergency operation tasks assigned to operators in power stations during incidents.

MERMOS considers the complex operation system: the operator team and its interaction with procedures and interface. In contrast to earlier (first generation) HRA methods, MERMOS does not use a generic error model. MERMOS requires the analyst to construct as many failure scenarios as possible, as brief descriptions of potential accidents (or at least of suboptimal operation), in a qualitative and structured way. Each scenario is then quantified, by assessing the probability that the various factors necessary for it to occur will happen together, and assuming that the scenario will persist until the point of failure. Put differently, we can say that the probability of each scenario is the probability that a particular context occurs (with a low probability) that will lead the system into a coherent but failing operation (with a high probability), because the inertia of the system operation is not stopped before an irreversible state (lack of reconfiguration). The overall probability that the task will fail is obtained by summing the probabilities for each failure scenario.

The operation of the system over time is described using the CICA concept. Because we are interested in a failure, only the CICAs that are necessary and sufficient to explain the failure are useful. (In a real operation, other CICAs may be superimposed.) It is the context that determines the conditions for CICAs arising. This context represents first a situation that has the structural characteristics that are the consequence of the initiator and the aggravating conditions that together define the human factors mission. But the context, beyond these structural aspects, is also a special context that allows the generation of the CICAs leading to the failure. The special context elements may be random minor events (failure of components such as indicators), the variability of the team (training, experience, operational style, etc.), and the possible history from the start of the incident (order of the events, other prior events, etc.). The probability that the scenario may arise is therefore the probability of the co-occurrence of the contextual elements and of the emergence of the CICAs. The probability of a failure is therefore the probability that the CICAs persist until the occurrence of the failure.

Our hypothesis is that, at a collective level, an operation is rational and consistent relative to the precise situation in which the operating system is embedded. To be able to operate effectively, the system is engaged in a strategic operation based on a hierarchization of the information and of the means to be used. The internal interactions of the collective of actors or at the human–machine interface will cause the emergence of the choice of parameters to be monitored, of the objectives to be attained, and of a priority in the means to be used. Once committed to this operation, the team at the center of the system will maintain, as far as possible, the adopted configuration and the targeted objectives, as long as a global reexamination is not appropriate. This could happen if a new event takes place or if the objectives have been attained, triggering a discontinuation of the actions. This operational inertia will show itself over time by a stabilization of the system configuration, of the means used, the parameters monitored, and the desired objectives. It means the system can resist the numerous demands placed on it in an environment of information overload and restrict itself to a given operation among the innumerable possibilities for action, which finally means that it can optimize its physical and cognitive resources.

In order to describe the stability of the operating system for a given duration, we therefore use the ad hoc concept of CICA to model the operation. Our model is as follows:

- The operative objectives of the operators may be explained at the collective level by the operating team in interaction with the interface and the procedures, even if an individual operator is unaware of the collective consensus or is opposed to it. In certain extreme cases the whole of the team may disagree with the operation, which it is effectively carrying out guided by the procedures. Even in this case we can attribute the objectives of the procedure to the system: we consider that the actions are determined cognitively by the knowledge stored in the procedures and that this is brought to bear in the situation by the operators who apply them. The description of the operation is therefore defined by the actions effectively carried out (or not carried out, as the case may be). At the other extreme, the operators may act contrary to or outside of the procedure by using their own cognitive resources. A symmetrical reasoning is applied to the modeling at the system level: the resulting operation will be attributed to the system in its entirety.
- To control the operation, the system necessarily has to manage the information at its disposal using its cognitive, material, and human resources in very complex situations. This management, which is more or less explicit and deliberate, requires a selection of the information to be processed, an allocation of resources and strategic priorities for the actions, and last but not least, resistance to external demands, thereby making savings in resources and ensuring the maintenance of the priorities for the action. To simplify, we can describe this management as resource configuration and strategic orientation in a given time period. The association of a configuration and an orientation is called a CICA.
- Definition: A CICA describes a transient inertia of the operation system determining a configuration of the resources and an orientation of the operation generating the appropriate actions, selecting and prioritizing the relevant information to be monitored, and setting aside the other actions that have low priority or are contrary to the orientation chosen, as well as the information that is less relevant in the context.

CICAs are rules the actors adopt to make sense of their collective safe action. Actors contribute to the emergence of these rules during the reconfiguration of the operation system. Safe operation can be seen as a succession of actions following the rules and a collective production of new rules. The safe regulation model integrates the dynamic of CICAs and reconfiguration and their sociological interpretation with the TRS.

THE SAFE REGULATION MODEL

We will reuse here the description of the dynamic of safe regulation from Le Bot (2008). The modeling using rules is particularly interesting to describe the effective organization of a nuclear plant, where the degree of prescription is extremely high, even if we should not confuse rules and procedures. We have seen that the notion of a rule that is observed by a collective has a strong resemblance to the CICAs, which we have developed for MERMOS to describe the orientation and configuration adopted by a group to decide which actions to perform and which not (Le Bot, 2004). We have therefore developed the safe regulation model to incorporate the TRS to explain collective behavior in a high-risk industrial environment. We have also tried to detail the model, according to our empirical observations for MERMOS on simulators, to include in it the safety requirements that must be satisfied by the organization of a nuclear plant. The essential particularity of the safe regulation model is to assume that the joint regulation, which builds the effective rules, is necessary in a situation (or just before the occurrence of the situation). This essential dimension of the model flows from the constraints inherent in the management of system safety in a situation.

To be safe, the organization has to permanently check the validity of the effective rules applied in a situation and, if necessary, provoke a rupture or discontinuation in the accomplishment of the actions, in order to produce new rules in a situation. These are built by negotiation between the control rules (existing prior to the situation) and the autonomous rules (produced in the course of the action). In the social theory of regulation, the reconstruction of the rules may be envisaged after the action, especially as the author is focused on the social rather than the technical issues. In a system at risk, the application of "bad" rules may call into question the safety of the system and therefore its existence. The social issues are overtaken by the technical issues as soon as the situation degrades or is at risk. For example, cultural or strategic issues can only weaken the safe regulation. The defenders of high reliability organizations have noted certain constants in the organizational characteristics of organizations: commitment of the leaders, redundancy, learning capacity, etc. (Sagan, 1993). In the same way, these ultra-safe systems all have organizational characteristics in terms of in-situation regulation. Our assumption is that these characteristics may be the framework of a reference model, the safe regulation model, that can be used to evaluate, by comparison, the reliability of an organization (or at least its lack of reliability). The safe regulation model is therefore an ideal model, serving as an evaluation reference.

What are the characteristics of in-situation joint regulation, more precisely? First, this model assumes that in organizations at risk, an action is undertaken by a collective within a distributed cognitive system. For example, the operating system in the control room of a nuclear plant, made up of a team of operators, interacts with its procedures and the remote control interface. This collective manages its action by the application of rules, this term being understood in its widest sense. If the rules applied most of the time are the procedures, the activity of the operators cannot be explained without taking into account the technical rules, the know-how of the operators, and the interpretation of the procedures themselves. But beyond the necessary adaptation of the prescribed task by the individuals, as ergonomists have amply shown, the work that we have done at EDF R&D with the orientation of human reliability has convinced us that a safe (or reliable) organization is explained at the level of the collective by the orientations and configurations, which it takes up in accordance with the situation, the CICAs, as shown in our retrospective analysis of the Three Mile Island accident (Le Bot, 2004).

In MERMOS, the CICAs describe this collective behavior and are therefore the effective rules adopted by the collective in order to manage the situation. CICAs are developed during rupture and reconfiguration phases and then applied during stabilization phases.

Effective Rules (CICAs in MERMOS)

The description of the operation is an account of the coherent functioning of the operating system observed in simulation and during incidents, emerging from the interactions between the parts of the system in the particular characteristics of the situations encountered. It is characteristic of the accidental operating system, which is a distributed cognitive and decision-making system.

At an instant t, the whole of the operation is described by several effective rules (CICAs) running in parallel. A change of situation, due to new events (change in the process, modification of material or human resources, new information in general, etc.) or due to the attainment of the objectives, may suggest a reexamination of the justification for the operation in progress. In this case a rupture of the operation is observed, that is to say, an interruption of the rules in progress, and a reorganization or reconfiguration of the system (new configurations, new orientations). Other effective rules are generated (some rules may persist). The observable symptoms of these ruptures are the stopping of the actions being performed, physical displacements of the personnel and/or their regrouping, multiple exchanges, changes of procedures, and recourse to new resources (field operators, teams on call). Any sudden change in the context, such as the crossing of physical thresholds, the triggering of important alarms, and expected feedback are, generally speaking, signals that announce ruptures.

The terms used to describe effective rules are, in general, terms expressing a transient attitude such as focusing, anticipation, prioritization, etc. (see Table 4.2). Elements that are observable in-situation and appropriate for effective rules are difficult to identify.

TABLE 4.2

Examples of Effective Rules

Suspending actions for an emergency shutdown

Focusing on restoring the water supply facilities for the steam generators

Delegating the monitoring of the auxiliary feedwater tank level for the steam generators

---

It is easier to deduce the rules by a retrospective analysis of the actions that were or were not executed than to observe them directly. However, symptoms may be identified through direct observation in the regularity of the geographical positioning of groups of operators (configuration of the resources), in their consultations (prioritized selection of information), and in their communications (orientation of the operation). Often the objectives of the procedure (control rules) organize the effective rules (attainment of the imposed reactor status, monitoring, etc.).

Effective rules emerge from interactions between actors, between actors and procedure, between actors and interface, and between the system and the situation. Their interruption requires a cognitive reexamination, which is often initiated by the redundant elements of the system: for example, in a nuclear plant, monitoring and inspection procedure, team supervisor, safety engineer, or crisis team. This difficulty of reexamining the operation in progress is at the heart of our definition of the failure. Failure occurs when there is no reconfiguration in parallel with the operation in progress, even though it was appropriate at the start of the situation: the inertia of the operation of the system is too great and the reorientation fails.

Individual errors must be recoverable and the collective and local organizations must be such that they can be recovered before the installation reaches an irreversibly degraded state. To include this capacity, the collective organization and the organization must be described in a dynamic way. The organization must be such that the actors can alternate between two essential organizational phases, which we call the stabilization phase and the rupture phase (Figure 4.1). Moreover, the organization must in parallel know how to implement a fourth regulation, the deferred joint regulation, which allows the cases experienced to be taken into account in a continuous and anticipative process of improvement. The deferred joint regulation is usually understood as the operational feedback.

The Stabilization Phase

During a stable operating phase, called the stabilization phase, the system follows the effective rules that it has set itself, allowing it to concentrate on its objectives and

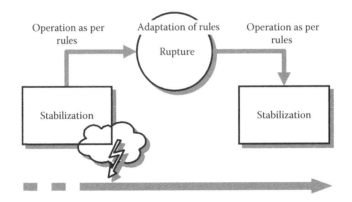

FIGURE 4.1 Phases of regulation.

to organize itself to attain them in accordance with the situation to be controlled. This stability is necessary in order to avoid the continual demands made by the dense flow of information (for instance, several hundred alarms in a nuclear power station control room). It allows the system to be robust. The guiding principle is not to look for any relevant information, but to concentrate on selected information. The operating system is focused on the objectives determined by the rules in order not to disperse its physical and cognitive resources. However, this organizational inertia, protecting the actors from unexpected demands, must be counterbalanced by permanent redundant surveillance (or monitoring). This surveillance constantly checks that the rules being applied are appropriate to the situation (for example, that the procedure in progress is adequate), in addition to the conventional monitoring for execution error recovery. The surveillance is undertaken, in general, by a manager of the team (in a nuclear plant, the supervisor or safety engineer). The control of the right application of the rules is undertaken by the actor and, implicitly or explicitly, also by the other members of the team.

The active rules can become inappropriate for two reasons. First, the objectives may have been reached, in compliance with the applied rules, and the operating system has to be reconfigured so that it has new effective rules, giving it the objectives, the diagnosis, and the strategy appropriate to the current situation. Second, the situation can evolve in an uncontrolled way, with either a new hazard arising or an error having been made, with an impact on the evolution of the situation, necessitating an organizational reconfiguration that is more than a simple recovery. In these two cases, the surveillance of the execution shall trigger a rupture of the operation so that the system reconfigures itself by starting up the in-situation joint regulation. This should provide it with new effective rules so that it can manage the process.

THE RUPTURE PHASE: IN-SITUATION JOINT REGULATION

"In-situation joint regulation" is the particular feature of the social theory of regulation that we are adapting for safe systems. It is implemented in rupture phases. We assume that the situations where a failure may occur are special situations in which the prescription, usually in the form of procedures but also by the training of the operators, is often not directly applicable (even with a prescription allowing the management of all situations with an extremely high probability). The operation system must therefore take into account the fact that there is a margin of error or non-coverage of the prescription. It must be able to adapt the control rules in order to ensure the safety of the operation. This lack of determination can in principle not be calculated; otherwise, it would be incorporated in the prescription, although a design error may produce the erroneous prescribed rules. In order to be safe, the operating system must therefore be able to adapt the rules in a situation by judging whether they are adequate or not. This necessary adaptation is in fact only an aspect of the necessary adaptation of the prescribed task to the reality of the experienced situation, in the sense that the ergonomics describe the activity, except that it is done collectively. This adaptation is made with three types of resources:

1. Direct knowledge of the prescription through the training of operators (which means, for example, that errors in an instruction may be identified).
2. Meta-knowledge of the prescription, enabling individuals to develop, in the situation, prescriptions based on the principles of the doctrine; this meta-knowledge relies on the culture of safety, as de Terssac (1992) has suggested.
3. The know-how and practices of the operators based on their experience and their on-the-job learning.

The operation system must continuously evaluate the prescription's adequacy in the situation, and this judgment, which by definition is human, may be erroneous in particular because of the complexity and the rarity of the situations in question. The result of this adaptation, called in-situation joint regulation, will be a negotiation between the prescribed rules and the autonomous rules, which we call effective rules.

As soon as a reconfiguration is triggered and the rupture phase has begun, the operation system must simultaneously select the control rules relevant to the situation (if they exist) and mobilize the culture of safety and the know-how of its operators. If it is necessary to implement this technical expertise, the collective will by cooperation decide on the elements to be dealt with by the control rules. A negotiation between the two types of rules should result in the negotiated rules, which should be confirmed by a person delegated to represent control regulation in a situation. These organizational subprocedures (selection of the procedures, cooperation, negotiation, validation) may be implicit or explicit, formal or informal. It should be noted that in the case of the accidental operation of a nuclear plant, the rupture phase may last from two minutes to several tens of minutes, and the stability phase may last several hours.

DEFERRED JOINT REGULATION

In-situation joint regulation may produce effective rules that go further than a simple interpretation of the control rules:

- The rules prescribed may have proved to be erroneous, inappropriate, too few, or too numerous.
- The know-how or experience of the people involved may have been shown to be insufficient or, on the contrary, have been able to compensate for the control rules.
- The effective rules may have been shown to be more effective than the existing control rules or autonomous rules.

These and other possible situations are inevitable, given the complexity of the system and the necessarily incomplete nature of the previously written control rules. The in-situation regulation is synonymous with its adaptability to chance events. A system never confronted by these phases of adaptation may one day prove to be too rigid in its normal organization, and it will not be able to deal with a situation without initiative. In addition, it is essential for the system to anticipate the recurrence of similar situations (or worse) by taking into account the feedback of these experienced situations, for example, when correcting its control rules.

However, it may be dangerous for the system to directly incorporate what it has learned in a situation in new formal, or informal, control rules. Once again, it is the complexity of the system that justifies the organization putting in place a minimum procedure to avoid, for example, "normalizations of deviance" in the sense explained by Vaughan (1996). The organization should therefore know how to learn from experienced events and furthermore even know how to generate these "experienced cases" at will, because as the organization learns, it will have less opportunity to test this adaptability in real time and to learn more about itself. In order to generate test situations deliberately but safely, the organizations at risk will have to use simulation, knowing that they cannot proceed by tests or errors. In its most common sense, a simulation of the accidental operation of nuclear power plants consists of using simulators of the control center where the operating system acts (simulators of a nuclear plant control room). Probabilistic safety assessments can be considered to be simulations as well. Other possibilities exist — in particular, the stories related by the actors recalling an event they have experienced (war stories, storytelling). In effect, narrated accounts of experienced events situate us in the universe where the incident took place, in some ways simulating or re-creating this event for the listener or reader of the account.

On-the-job learning is direct in-situation feedback: the collective here plays an essential role through the various mechanisms of mutual assistance, cooperation, etc., meaning that the less experienced can have access to more experience under the supervision of their more skilled colleagues. On-the-job learning also means that the group may accumulate collective experience in the situations that it overcomes, particularly in consolidating this collective working. On-the-job learning has immediate consequences. However, feedback requires that what has been experienced is remembered, by both the actors and the technical setup (recordings of the parameters, cameras, etc.). The volatility of this memorizing means that analysis of the situations has to be done as soon as possible.

THE CAUSE OF FAILURE ACCORDING TO THE SAFE REGULATION MODEL: THE MAINTENANCE OF INEFFECTIVE AND INADEQUATE RULES

According to MERMOS, the empirical cause of failure is the maintenance of inappropriate effective rules (CICAs) until an irreversible situation is reached (Le Bot, 2004). A necessary condition for failure is therefore the absence of reconfiguration of the operating system to interrupt these inappropriate effective rules and to produce new ones that can manage the accident. Transposed to the safe regulation model, the explanation is that inappropriate effective rules govern the behavior of the collective until a point of no return is reached; that is, that a new in-situation joint regulation has not been triggered in time. By deduction, working backwards in time, the causes of the failure are either an undetected change in the situation, which makes the current effective rules obsolete, or a generation of inappropriate rules at the beginning, which have not been recognized as inadequate. In every case the triggering of a new regulation is faulty. The use of the safe regulation model means that possible causes can be found upstream. Failure of surveillance can have several causes; for instance:

- The effective rules that have been produced may have organized ineffective surveillance. For example, the team may have decided that it is unnecessary to call the safety engineer, whereas the latter would have been the only one able to detect the abnormality of the situation.
- The organizational resources are insufficient for the surveillance. For example, the safety engineer or the operations shift manager who replaces the former in his or her absence is occupied with another task.
- The surveillance is entrusted to a single individual who commits an error. For example, the safety engineer, being alone in carrying out the surveillance, does not detect an abnormal change in the situation.
- The surveillance is undertaken in a non-independent fashion. For example, the safety engineer is involved in the strategy of the team and is therefore not in a position to spot any inadequacy with respect to the situation.

Failure of in-situation joint regulation may have causes that are even more complex. Some examples:

- One of the "cogs" in the organization is missing; either it is not planned or does not work because of a lack of resources. For example, the validation of the effective rules does not operate for lack of sufficient delegation to someone in the control room who can accept responsibility for the decision.
- The operation of the collective does not allow joint regulation. A breakdown in communication between the team members and/or the team and its procedures inhibits dialogue, so that the various regulations are expressed. For example, an experienced operator does not dare draw attention to his or her experience for fear of contradicting the line manager.
- One of the regulations does not function. For example, the situation in progress is covered by neither the procedures nor the principles of safety culture. The operators are therefore forced to rely only on their know-how and their practices in order to invent an operation in a complex risk situation.

It is even possible to demonstrate deeper causes with the failure of deferred joint regulation (operational feedback). For example, the absence of after-the-event analysis of the rules that have been used during an event may lead to an inability to spot the inadequacy of the control rules. The rules may have been compensated by the skill of the operators present for this event, but will not necessarily be compensated at the next similar event. Conversely, if an autonomous rule is used without knowing the consequences for an event, and if it is not reexamined and confirmed after the event (and therefore incorporated into control regulation), it may be adopted to manage another event when in fact that would be dangerous.

DEFINITION OF ERROR WITH THE SAFE REGULATION MODEL

Woods has convincingly argued that, "Attributing error to the actions of some persons, team, or organization is fundamentally a social and psychological process and not an objective, technical one" (Woods et al., 1994).

In practice, it can be agreed that error is a behavior that deviates from a social reference. The psychological side of error concerns the causes and the modes of errors but not intrinsically the erroneous nature of behavior that we can restrict to social aspects. Then the problem is to define the social reference that can be used to assess the erroneous quality of behavior.

More often, people use their own social reference to judge behavior. Consensus and objectivity are then impossible, because the reference is subjective. With the safe regulation model, we can show that actors themselves design their own social reference in a situation, leading to effective rules that actors share in a collective to reach their objectives. We can then state that an error occurs when behavior is deviating from the effective rules that are followed by a collective in a situation.

If that behavior is restricted to one actor, we called it an individual human error. It occurs when the actor's behavior does not follow the effective rules of the group he or she belongs to. If somebody deliberately does not follow the group's rules, we have to verify whether the person considers himself or herself a member of the collective and acts as such before we qualify the behavior as an error. A group, a collective, can behave erroneously when its collective functioning is not adequate. For example, poor communication can lead a group to act against its own effective rules. An organization can lead a group to act erroneously by preventing its access to adequate resources and means for its good collective functioning.

The obvious problem with our definition is understanding what the effective rules are that a group has followed during an accident, because effective rules are not permanent and are fitted to each particular situation. We consider that we can find these rules by induction. We analyze the operation that has been performed, or avoided, and use interviews of the actors if possible. The counterpart and the positive issue of that definition is that it avoids many of the problems that come from a classic error reference that is external to the actors. For example, if following the procedure is included in the non-error category, the special situation where an operator has to apply a flawed procedure with a bug is paradoxical, because if he follows the procedure the action fails, and if he does not follow the procedure, it is an error. With our definition, it depends how the collective in the situation has coped with the bug in the procedure, and which effective rules about the procedure have been produced in the situation. If the collective makes a wrong diagnosis and decides to follow the procedure, the operator who follows the flawed procedure does not make an individual error.

ABOUT HUMAN ERRORS AND HUMAN FAILURE
FROM AN HRA POINT OF VIEW

From our experience, individual human error is not predominant, nor does it explain the failures. In each failure scenario, the system is solid and redundant enough for the consequences of an individual error to be contained. Such an isolated error cannot lead to failure given the barriers and redundancies in place. For example, an operator's procedural error may be corrected by another operator or the supervisor who observes anomalies in the procedures. If the whole team is involved in an erroneous operation on the basis of an inadequate procedure, this operation cannot be taken to be a simple error of procedure implementation; it can arise only from a conscious operation of the team, which has decided on this course of action for good reasons due to the particular characteristics of the situation. This action may have been originally initiated by an individual error of one of the team members, but this cause is not sufficient. So we consider this error, which contributes to the generation of the final operation, as a feature of a special situation (or context). This situation or context should be analyzed in as much detail as possible so as to explain that the whole system may be involved in this erroneous operation.

At a collective level, the operation is rational and consistent relative to the precise situation in which the operating system is embedded. The description of the operation at the system level makes individual errors irrelevant to explain the failure. Incoherence arising from individual error is treated at the collective level by redundant error recovery systems, whether they are material (auxiliary redundant systems), procedural (redundant procedures), or organizational (safety engineer, supervisor). However, in terms of safety, this individual error exists, even if it cannot generate the global failure of the system. It is taken into account in our model as an element contributing to a special situation.

A SHORT EXAMPLE OF REGULATION AND ERRORS

Let us explain the safe regulation model with an imaginary and simplified example from daily life in a risky system: the roads in France. Consider a little road, R, that merges onto a highway, H, that is very often busy. To merge, the highway drivers on R must stop and give the priority to the drivers moving on the highway. Imagine an accident that involved an experienced driver, E, coming from the little road, R, and a young driver, Y, on the highway, H, that crashed into E's car.

Is E at fault for erroneously denying the priority to Y? Or is Y guilty because he or she could have made an error from lack of experience; for example, by driving too fast? (Without considering legal aspects, we are trying to understand how to explain the accident from an organizational perspective.) A retrospective analysis of the event may show the following facts:

- Two hours before the event there is a traffic jam on the highway at time t_0. The drivers on H therefore slow down; after a while drivers on R cannot merge onto the highway because they have to stop, waiting for enough space in the stream of cars on H to merge in. Because the speed on R becomes very low, thirty minutes after the beginning of the traffic jam, an impatient driver on R does not respect the priority and engages his car on H. Drivers on H, understanding that drivers on R have to wait for a very long time before there may be a break in the stream on H, sometimes stop to let drivers from R merge onto the highway. After twenty minutes, drivers from H and drivers from R stop alternately to let each other run on H.

- After two hours, the traffic jam is decreasing and the speed on H is increasing. Driver E on R, thinking that the cars from H will allow him to merge onto H, because he has observed the previous conditions, therefore tries to enter onto H. There, driver Y, who may be moving fast (but under the legal limit) because there is no more traffic jam, cannot stop and crashes into E's car.

With the safe regulation model we can analyze the accident by recognizing four consecutive phases:

1. From t_0 to $t_0 + 30$ minutes: This is a stabilization phase. The rules are the control rules, and drivers respect the priority rules.
2. At $t_0 + 30$ minutes: This is a rupture phase. A negotiation is engaged in between control rules (supported by H's drivers) and autonomous rules (supported by R's drivers). Experienced R drivers know on the one hand that if they do nothing in the situation, they will have to wait a very long time. On the other hand, they also know that in this situation the low speed allows them to try to merge the highway without respecting the priority, and furthermore that H's driver will understand. A new effective rule emerges from the interactions: drivers from H and drivers from R stop and continue on H alternately.
3. From $t_0 + 30$ minutes to $t_0 + 2$ hours: The rule has stabilized the operation and the two streams from H and R are merging on H in a smooth way.
4. After $t_0 + 2$ hours: The situation has changed but the rupture is partial rather than complete. Drivers on H, seeing that traffic on H eases, begin to give up the effective rule adopted during the third phase. On R the stream is the same, and driver E, thinking that the rule is still being followed, moves onto H. Driver Y, ignoring the effective rule, is running fast and cannot stop.

From the safe regulation model point of view we can state the following issues:

- Neither E nor Y made any errors because they were following the effective rules adopted by the collective to which they belonged.
- The system composed of the cars, the drivers, and the two roads is not safe because the regulation misses delegation of control regulation and the surveillance of the adequacy of effective rules to the situation, and because the collective functioning is poor (i.e., the actors are communicating only visually, with little information exchange).

The lack of a delegation of the control regulation means that there is nobody (for instance, a police officer) who can validate the effective rules that have been elaborated by the collective. The lack of surveillance means that nobody is formally in charge to verify whether the effective rules fit the situation. It is therefore difficult for the drivers from R to know exactly when the effective rules that applied during the traffic jam are no longer valid. The lack of communication breaks the collective into two groups (drivers on H and drivers on R), who then begin two adopt their own rules and break their cooperation.

CONCLUSION

The explanation of effective rules using the safe regulation model allows us to develop the theoretical basis of the empirical model developed for MERMOS and opens up several parallel perspectives of methodological expansions. One of the first benefits may be that CICAs, as described in MERMOS, are better understood in terms of effective rules. But above all, the sociological interpretation of the behavior of the operating team, in interaction with its procedures and its interface, appears promising. This interpretation shows the impact of upstream organizational factors on the safety of the accidental operation of reactors. The enriching of the safe regulation model and its link with MERMOS will be carried out according to two themes: the retrospective analyses of actual events on the one hand, and an already produced synthesis of the analyses of MERMOS itself on the other. Finally, an external validation could be done by testing the safe regulation model on organizations other than nuclear reactors. The first analysis we did with that objective was published in Le Bot (2008). The safe regulation model provided a good filter to look at the weaknesses of a risky system very different from a nuclear plant — the oncology department of a Scottish hospital. The next step is to refine the model to better support HRA and to use it to expand MERMOS.

REFERENCES

de Terssac, G. (1992). *Autonomie dans le travail*. Paris: Presses Universitaires de France.
Le Bot, P., Bieder, C., Desmares, E., Cara, F. & Bonnet, J.-F. (1999). *MERMOS, a second generation HRA method: What it does and doesn't do*. Paper presented at the PSA '99, Washington, DC.
Le Bot, P. (2004). Human reliability data, human error and accident models illustration. *Reliability Engineering & Safety Systems, 83*(2).
Le Bot, P. (2008). Analysis of the Scottish case. In E. Hollnagel, C. P. Nemeth & S. Dekker (Eds.), *Remaining sensitive to the possibility of failure* (Vol. 1). Aldershot, UK: Ashgate.
Nathanael, D. & Marmaras, N. (2008). Work practices and prescriptions: A key issue for organizational resilience. In E. Hollnagel, C. P. Nemeth & S. Dekker (Eds.), *Remaining sensitive to the possibility of failure* (Vol. 1): Aldershot, UK: Ashgate.
Reynaud, J. D. (1989). *Les règles du jeu*. Paris: Armand Colin.
Sagan, S. D. (1993). *The limits of safety: Organizations, accidents, and nuclear weapons*. Princeton, NJ: Princeton University Press.
Vaughan, D. (1996). *The Challenger launch decision: Risky technology, culture, and deviance at NASA*. Chicago: University of Chicago Press.
Woods, D. D., Johannesen, L., Cook, R. & Sarter, N. (1994). *Behind human error: Cognitive systems, computers, and hindsight*. CSERIAC SOAR Report 94-01. Crew Systems Ergonomics Information Analysis Center, Wright-Patterson Air Force Base, OH.

5 Workplace Safety Climate, Human Performance, and Safety

Hiroyuki Shibaike and Hirokazu Fukui

CONTENTS

SAFETY CULTURE AT THE WORKPLACE

In recent years, the term "safety culture" has gained widespread use not only in the nuclear power industry but also in all other industries. Based on lessons learned from the Chernobyl nuclear power plant disaster in April 1986, this concept has been advocated by the International Atomic Energy Agency (IAEA, 1991). In 1991, the IAEA defined safety culture in the International Nuclear Safety Advisory Group's report as "an aggregation of the traits and stances of organizations and individuals seeking to give nuclear plant safety the attention it deserves and to ensure safety" (IAEA, 1991).

Furthermore, the international nuclear energy watchdog publicized the universal characteristics of safety culture and indicators for evaluation by a broad audience, from governments to individual citizens (IAEA, 1991; IAEA, 1996). This concept is notable in that it has helped the nuclear power industry, which has made tremendous efforts to raise public confidence with regard to its hardware (technology and buildings), to recognize once again the importance of organizations and individuals. Regrettably, however, the proposed safety culture indicators fail to show how they are related to one another and what impacts they have on the safety consciousness and behavior of individuals in charge of handling nuclear plants.

On the other hand, *climate* is a synonym, albeit only partially, for *culture*. Human beings have been living in the natural environment and have created lifestyles befitting the natural environment. This is the culture. Human beings always stand between the natural environment and the culture. Climate represents a person's recognition of the natural environment, whereas culture may be said to be a generic term for what people have built for their lives, based on climate. Accordingly, activities that do not harmonize with the climate will die out and a unique culture that is harmonious with the climate will be created. There is a causal relationship between climate and culture.

It may be a natural consequence of this situation that the term "safety culture" is closely associated in our minds with the term "safety climate." Although nuclear power plants are operated by organizations, it is the individuals within the organization who are directly handling the technology and equipment. Clearly, individual behavior is closely connected with safety. In other words, ensuring that individuals consider safety and safe behavior is one of the ultimate goals in preventing human errors, and it is an organizational environment that guides people to that goal. Thus, a safety climate can be defined as an organizational environment that guides its members to consider safety and safe behavior (Fukui, 2001). Whereas safety culture encompasses the traits of an organization and of individuals, safety climate focuses not on individuals but on an organizational environment. Safety culture is built on a safety climate. This perspective of a safety climate is needed to cultivate a safety culture.

RATING SCALE

A safety climate rating scale was developed as follows. Based on the IAEA's ASCOT (Assessment of Safety Culture in Organizations Team) Guidelines and on lessons learned from case studies, approximately one hundred rating items on a scale of 1 to 5 were prepared. A questionnaire was then sent out to about twelve hundred personnel at three nuclear power plants. Based on the data derived from the replies to this questionnaire, a number of factors were identified, of which six were screened out. For each of these, five relevant statements were chosen and used to construct the safety climate statements shown in Table 5.1. For each statement, respondents were then asked to choose a response in accordance with the following five-point scale:

1. Disagree
2. Disagree somewhat
3. Neutral
4. Agree somewhat
5. Agree

TABLE 5.1
Composition of the Safety Climate Rating Scale

Factors	Relevant Statements
The organizational attitude for safety	1. Plant executives agree with the organization's stance and efforts toward safety.
	2. Plant executives' enthusiasm and determination toward safety efforts affect members of the organization.
	3. Plant executives discuss safety issues.
	4. Plant executives examine safety by conducting walk-through inspections of the workplace.
	5. Your immediate superior approves of his staff's stance and efforts toward safety.
Attitude of immediate superior	1. Your immediate superior considers the working environment to ensure his staff can work satisfactorily.
	2. Your immediate superior assigns tasks to his staff based on a full understanding of their abilities and conditions under which they work.
	3. Your immediate superior makes efforts to ensure that his staff who have contributed to the enhancement of safety are recognized by the power plant management.
	4. In your workplace, you are given a convincing explanation of the details of your job.
	5. It is clear who is responsible for each task to be performed in the workplace.
Workplace safety motivation	1. Your workplace atmosphere allows employees to sincerely discuss difficult safety issues.
	2. In your workplace, employees discuss accident and safety issues frankly.
	3. In your workplace, employees discuss near-miss incident experiences.
	4. In your workplace, employees' opinions and ideas designed to ensure safety are adopted.
	5. In your workplace, employees make proposals for improvements that will help enhance safety and workability.
Safety-conscious behaviors	1. Before starting work, your fellow workers check that safety in the workplace is ensured.
	2. Your fellow workers first check the safety of a more efficient work method they have devised before carrying it out.
	3. Your fellow workers check in advance for any hazards existing in a work area.
	4. Your fellow workers pay attention to safety by carrying out an on-site inspection for any safety problems.
	5. Your fellow workers give priority to safety even when very busy with their work.

(continued)

TABLE 5.1 (CONTINUED)

Composition of the Safety Climate Rating Scale

Factors	Relevant Statements
Morals	1. Your fellow workers sometimes bypass the prescribed procedures or rules for reasons just for the occasion (a negative element).
	2. Your workplace sometimes chooses an easy method instead of observing troublesome regulations or rules (a negative element).
	3. In your workplace, most of your fellow workers are self-confident as professionals who assume social responsibility for their job.
	4. In your workplace, most of your fellow workers consider it important to work from general citizens' point of view.
	5. In your workplace, some of your fellow workers express personal feelings and interests at work (a negative element).
Confidence in knowledge/skills	1. You are equipped with the knowledge and skills needed to perform your duties.
	2. You are confident in responding calmly to any accident.
	3. You have the knowledge and skills needed to ensure safety.
	4. You are well adapted to the progress of facilities and equipment.
	5. You have detailed knowledge of the duties of your fellow workers.

The sums of the numerical value of the responses were calculated, so that each factor received a score between 5 and 25.

The ratings for each workplace were returned to each workplace after comparison with the results of the previous survey and with the standard value. Based on the rating results, the personnel discussed the strengths and weaknesses of the workplace and determined a policy for improvements. The flow of feedback is shown in Figure 5.1. Workplaces that had a comparatively high rating were subjected to a field survey to identify best practices (see Chapter 10).

SURVEY RESULTS

Safety climate surveys are conducted about every second year, targeting the managers of engineering-related sections and their subordinates at nuclear power plants and other such workplaces. We have already conducted six such surveys at nuclear power plants. We sometimes conduct a similar survey at higher organizational levels such as the head office, in response to requests from such workplaces.

Figure 5.2 shows assessment results from surveys in FY2005 and FY2006 conducted at a workplace in a nuclear power plant. This workplace had ten employees. The figure shows improvements in *organizational attitude for safety, attitude of immediate superior, safety conscious behaviors, morals, confidence in knowledge/skills, teamwork,* and *mental well-being,* but a decline in *communication.*

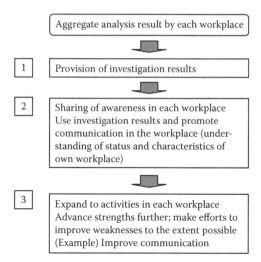

FIGURE 5.1 Feedback flow of investigation results.

Results like these are fed back to members of the given workplace for discussions within the team. Employees define new goals for their efforts based on their deepened awareness of the characteristics and weaknesses of their workplace. They do not spare efforts to achieve these goals because such goals are formulated by themselves after discussions. Periodical implementations of these surveys supports the PDCA (plan–do–check–act) cycles of improvement activities for the establishment of an excellent safety climate.

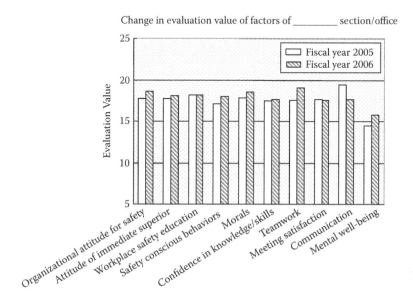

FIGURE 5.2 Aggregate analysis result of workplace.

MANAGEMENT AND LEADERSHIP

PM Theory

The principles of PM theory are described in detail in Chapter 11, and we will therefore only provide an outline of the theory here. (Additional information can be found in Misumi, 2001.) In the field of group dynamics, the functions of a group are often classified in two categories. One is a target performance function and the other is a group maintenance function. The target performance function is the action intended to help a group achieve a given target or solve a given problem and is referred to as the P function, where P stands for *performance*. The group maintenance function is the action designed to properly maintain unity and human relationships among members of a group, and is referred to as the M function, where M stands for *maintenance*.

However, the functions of an actual group are often closely connected with the behavior of its leader. In the leadership PM theory, therefore, leader behavior that seeks to accelerate the P function of the group is referred to as P behavior and leader behavior that intends to promote the M function of the group is referred to as M behavior. As shown in Figure 5.3, when the P behavior and the M behavior are combined on two dimensions, leadership can be classified into four types. The *PM* type indicates that the leader is strong both in the P behavior and in the M behavior. The *Pm* type indicates a leader who is strong in the P behavior but weak in the M behavior. The *pM* type indicates a leader who is weak in the P behavior but strong in the M behavior. The *pm* type indicates a leader who is weak in both the P behavior and the M behavior. This approach to studying the motivation of group members and a group's productivity has yielded the finding that members working under PM-type leaders have the highest level of motivation and productivity, whereas members working under pm-type leaders have the lowest level of motivation and productivity.

Survey Method

For leadership assessment, we conduct questionnaire surveys to have each subordinate of the target manager assess the manifestation of leadership by the manager in terms of the aforementioned PM theory. The questionnaire comprises ten questions

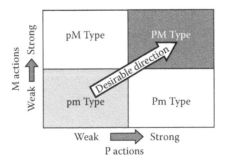

FIGURE 5.3   Four types of leadership (PM Theory).

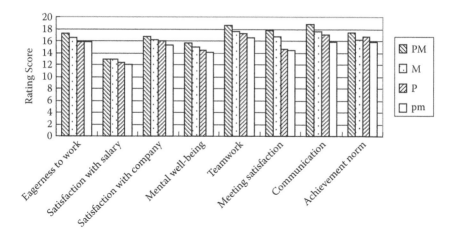

FIGURE 5.4   Correlation between leadership types and morale (for an NPP manager).

about P leadership and ten questions about M leadership. One of the questions about P leadership is formulated as follows: "Does your boss ask you to report progress of work?" Respondents choose a response from a five-point scale.

As a result of questionnaire surveys, the managers are classified into four categories by their leadership characteristics: PM type, Pm type, pM type, and pm type. The results are fed back to the managers, but care is taken not to allow the results to be traced back to individual respondents. These surveys make managers more deeply aware of their shortfalls in the manifestation of leadership and encourage them to improve their management skills based on this awareness.

We conduct such surveys in various styles according to requests from client organizations. For example, PM surveys could be conducted three times in a period of about six months, or a PM survey could be followed up by a half-day leadership seminar.

SURVEY RESULTS

Figure 5.4 provides an example of results from leadership surveys that we conducted in 1995 and 1996. The figure shows the relationship between the leadership of section managers at nuclear power plants and the morale of team members. With each factor, workplaces with a PM-type manager are evaluated most highly.

A comparison between the first and third PM surveys on 62 managers revealed that 24 managers maintained PM-type leadership, 14 managers advanced from non-PM-type leadership to PM-type leadership, 22 managers remained stuck at non-PM-type leadership, and 2 managers fell from PM-type leadership. This demonstrated the usefulness of our leadership surveys and seminars because more than a half of the target managers have shown advancement.

Like the safety climate surveys described earlier, it is important that we conduct these leadership surveys periodically.

ROOT CAUSE ANALYSIS

It is crucial to absorb lessons learned from problems caused by human errors to prevent further human errors. In order to do this, we must have a correct understanding of events and be able to identify factors that contributed to human errors. At Kansai Electric Power, if a problem or an occupational accident occurs due to human errors at a nuclear power plant, the plant general manager and higher organization determine the necessity of analyzing the human errors. If it is decided to do so, an analysis team composed of plant personnel is set up to conduct the analysis.

An effective analysis hinges on how accurately events can be understood. To ensure that, as many stakeholders as possible are interviewed. In each interview, it is essential to convince the interviewees that the purpose of the analysis is not to blame individuals but to prevent a repeat of similar problems and thereby to encourage them to tell the facts.

Because the Japanese government requires electric utilities to analyze underlying causes in accordance with the regulatory guidelines, it is crucial to examine organizational factors in particular, and analyses of facts are therefore even more important.

ANALYSIS METHOD

The Kansai Electric Power Company developed, on its own, "the human error fault tree diagram technique" in order to decrease the problems caused by human error in field operations at nuclear power plants. In addition, we are carrying out other activities designed to share lessons learned from troubles that have occurred in our own workplaces so that other plants can learn from these lessons in their workplaces.

One principle on which the human error fault tree diagram analysis is based is that the blame should not be placed on any party that was involved in the human error. The basis of this technique is to rectify the tendencies to put the blame on personal factors by incorporating that party's information processing systems and working environment factors in the analytical technique systematically.

For a conceptual model, the analytical technique is based on the error occurrence process model presented by Rasmussen (1983) and Swain & Guttmann (1983). The technique designates top events as human errors, intermediate events as people's information processing phases, and lower events as internal and external factors. Internal factors are organized according to the classification based on Hashimoto's (1984) consciousness phase theory, whereas external factors are arranged using classification based on the four Ms (man, machine, media, and management). Figure 5.5 shows the basic structure of the human error fault tree diagram.

Table 5.2 explains the consciousness phases. To prevent human factor errors, we must adopt measures based on an understanding of the level of consciousness at which people tend to commit an error. By interviewing persons who committed an error, for example, we identified the level of consciousness at which they committed the error, using a five-point scale of consciousness that stretched from Phase 0 to Phase IV. Normally, people are in Phase II when working but shift into Phase III

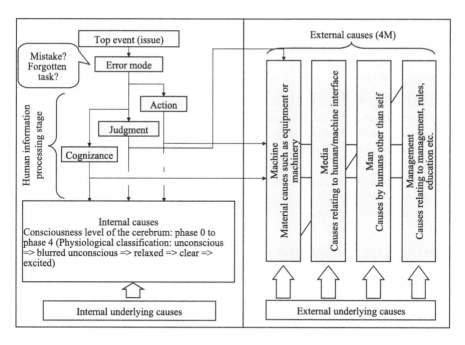

FIGURE 5.5 Basic structure of human error fault tree diagram.

when faced with a task that requires extra attention. However, they must come back to Phase II after a while because it is difficult to stay in Phase III for a long time. Veteran workers are skilled at switching between Phases II and III. Phase I is the typical state of consciousness when a worker is continuously involved in monotonous tasks. Phase IV is the typical state of consciousness when one is pressed for time.

TABLE 5.2
The Consciousness Phases

Phase	Mode of Consciousness	Function of Attention	Physical Condition	Reliability
0	Unconscious or lost consciousness	Nil	Asleep or brain attack	0
I	Subnormal consciousness	Careless	Fatigue, monotonous, sleepy, or intoxicated	Below 0.9
II	Normal — relaxed	Introverted	Living calmly, at rest or normally working	0.99–0.99999
III	Normal — clear	Forward-looking	Vigorously engaging in an activity	Over 0.999999
IV	Hypernormal — excited	Obsessed with one single idea	Excited or in panic	Below 0.9

Workers must avoid being in either Phase I or Phase IV because these are the states in which the reliability of human actions tends to decrease.

ANALYSIS RESULTS

This analytical technique has been developed and employed as an approach that is easy to understand, so that field staff can use it properly. The technique also has an educational aspect because it allows us to make a deeper study of the actual conditions of human errors through analysis. Since its development in 1987, this technique has been used to analyze approximately fifty cases. Because the japanese government required electric power companies to analyze underlying causes in accordance with the regulatory guidelines in December 2007, the analytical technique has recently been employed by each nuclear power plant to deal with one to four cases a year.

RAISING SAFETY AWARENESS

We produce single-point advice sheets to feed back lessons learned from plant accidents and other trouble to the power company and cooperating company employees, so that we may help the growth of their safety awareness. These sheets describe causes of trouble and actions taken against them, using easily understandable illustrations. About one hundred seventy such sheets have been produced since 1988. We are not producing these sheets for every accident or trouble; rather, these sheets are meant to address accidents and trouble that strongly teach us a lesson, and they are produced as soon as possible after each such event.

Every time we produce such a sheet, we distribute it to power companies and other cooperating companies. When we have produced about one hundred sheets, we compile them into a booklet for distribution (Figure 5.6). These sheets contribute to increased safety awareness at workplaces because they are used extensively at pre-work briefings on precautions, case study meetings, and so forth.

FIGURE 5.6   Creation and distribution of one-point advice sheets.

SUMMARY

The importance of safety culture has been emphasized for a long time, notably since the Chernobyl disaster in the former Soviet Union in 1986. In Japan, there has been a series of accidents indicating a decline in safety culture, such as the 1999 JCO criticality accident, the 2002 reactor vessel shroud cracking concealment scandal at the Tokyo Electric Power Company's nuclear power plant, and the 2004 Mihama-3 secondary system piping fracture accident.

These accidents were not just the consequences of hardware problems but also resulted from complex interrelated social and organizational factors. The present conditions can no longer be fully addressed by the human–machine interface approach alone. Under these circumstances, the workers' sense of values, their ways of thinking, their behavioral patterns and habits may hold the key to preventing accidents. This is what we call the safety culture of the organization to which people belong. Safety culture is created within the context of relationships among people. The activities described in this chapter represent efforts to create a highly trustful working environment within the context of relationships among members of an organization.

REFERENCES

Fukui, H. (2001). Genshiryoku Hatsudensho ni Okeru Anzen Fudo (Safety climate at nuclear power plants). *Electrical Review*, *86*(5), 31–35.

Hashimoto, K. (1984). *Anzen Ningen Kogaku (Safety human engineering)*. Tokyo: Japan Industrial Safety and Health Association.

International Atomic Energy Agency (IAEA). (1996). ASCOT guidelines, revised 1996 edition (IAEA-TECDOC-860).

International Nuclear Safety Advisory Group (INSAG). (1991). Safety culture (IAEA Safety Series No.75- INSAG-4).

Misumi, J. (2001). *Ridashippu to Anzen no Kagaku* (Leadership and safety science). Kyoto: Nakanishiya Shuppan.

Rasmussen, J. (1983). Skills, rules, knowledge: Signals, signs, and symbols and other distinctions in human performance models. *IEEE Trans. on Systems, Man, and Cybernetics (SMC)*, *13*, 257–267.

Swain, A. D. & Guttmann, H. E. (1983). *Handbook of human reliability analysis with emphasis on nuclear power plant applications* (NUREG/CR-1278). Washington, DC: U.S. Nuclear Regulatory Commission.

6 A Management Process for Strategic Error Prevention

Yuko Hirotsu

CONTENTS

INTRODUCTION

This work is concerned with the development of methods for analyzing and utilizing data from events due to human errors (i.e., inappropriate or inadequate human actions) for proactive accident prevention. Whenever an accident or a serious incident occurs at a plant, a thorough investigation is conducted to understand the causes, so that corrective actions can be developed to minimize the likelihood of its recurrence. In some cases, human errors in the field may directly be the cause of an event. In other cases, problems may have been embedded in equipment at the design phase and have been overlooked for a long time (see the description of the Mihama-2 accident in Chapter 2). In other words, it may be assumed that human performance to some extent is involved in any kind of event. Yet human performance is not a simple matter. Problems in human actions are often influenced by combinations of organizational factors, such as work planning and training, and work environment factors, such as the on-site situation. To develop effective countermeasures, it is therefore essential to clarify problems and causal factors of events from the aspects of human factors. However, relationships among the causal factors have become more and more complicated because organizations and facilities get increasingly complex, and it has become increasingly difficult to identify fundamental causes.

On the other hand, recent years have seen organizations beginning to share information on minor events attributed to human error (human error events) with the aim of preventing recurrence of similar events. Sharing information on human error events and systematically analyzing them to find their causes leads to clarification of problems in management and work climate. If it becomes possible to cope appropriately with these problems and to fundamentally improve work management method and work process, it will make it possible to prevent not only the similar types of errors but also different types of errors, and thus serious accidents. However, there are cases where such information is not fully utilized.

Various analysis methods for human error events have been devised and are utilized in industries such as nuclear power generation and chemical processing; for example, Barnes & Haagensen (2001) and CCPS (Center for Chemical Process Safety, 1994). The Central Research Institute of Electric Power Industry (CRIEPI) in Japan developed a human error analysis method, J-HPES (a Japanese version of the human performance enhancement system), in 1990 (Takano, Sawayanagi, & Kabetani, 1994), and has also been conducting analysis of human error events at nuclear power plants. Moreover, referring to the analysis results, CRIEPI has published posters and educational materials to further the prevention of human errors. However, a proven method for comprehensive and systematic analysis, including the consideration of organizational factors, monitoring of overall trends, and utilization of the analyzed results, has so far not been available.

This chapter presents methods for analyzing and utilizing human error event data that have been accumulated in an organization, in order to identify and address the latent weakness of an organization. We developed a basic concept of human error event analysis and incorporated it in the J-HPES in order to find commonalities among human error events. As a result, these analysis methods enable investigators to clarify organizational factors such as work control process and inadequate practices of workers. We also developed a management process based on event analysis results using these methods as a further study. The methods discussed in this chapter were tested mainly in the Japanese nuclear power industry (Hirotsu et al., 2006).

In the following, a basic framework for human error event analysis is first introduced. Then, the modified J-HPES, an analysis method for human error events, is described. Next, a method for identifying the commonalities among analysis results of human error events is explained. Finally, a management process for strategic error prevention is proposed.

BASIC FRAMEWORK FOR EVENT ANALYSIS

This section introduces a framework to assist event investigators in their examination of various causal factors of human error events. The J-HPES was developed by fully modifying the HPES method devised by the U.S. Institute of Nuclear Power

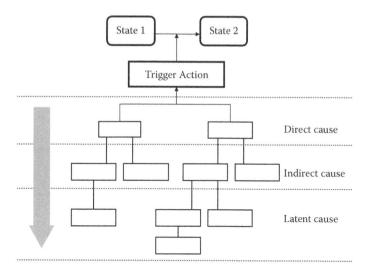

FIGURE 6.1 Causal relation chart (J-HPES).

Operations, so that it was adapted to a Japanese environment (Takano, Sawayanagi, & Kabetani, 1994). Developed as a remedy-oriented system for systematically analyzing and evaluating human-related events occurring at nuclear power plants, this method aims in particular at identifying causal factors and deriving proposals for specific hierarchical countermeasures. The procedure of the J-HPES comprises four stages:

1. Correct understanding of events
2. Circumstantial analysis (gathering human factor data)
3. Causal analysis
4. Proposing countermeasures

The causal analysis step (Step 3) is applied to each trigger action (defined as a human action contributing directly to an abnormal change of machinery state in an event). The approach applies the modified fault tree method to initiate a search reaching down to the ultimate underlying causal factors. This causal relation chart (Figure 6.1) clarifies the direct causal factors that have induced the trigger action, indirect causes that have contributed to the direct causal factors, and latent causes that have contributed to the indirect causes. The procedure of J-HPES is similar to the "why–because analysis" (Paul-Stüve, 2005). The J-HPES, however, especially focuses on the human actions that trigger an event and places emphasis on human factors when searching for the underlying causes of the actions.

The basic idea of the causal analysis is to ask the question "why" repeatedly, starting from "why does the trigger action occur." Thirteen categories of causal factors, such as work practice and work verification, are given as references for examining possible causal factors. An advantage of this method is that there are only a few rules and conventions, so that investigators can propose causal factors on their own. However, there is a concern that analysis results may vary too much because the choice of potential factors to examine and the decision of when to stop asking the

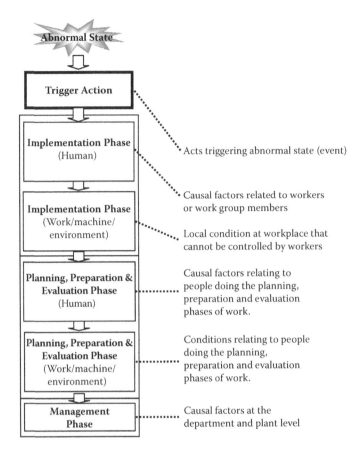

FIGURE 6.2   Basic framework for human error event analysis.

"why" questions are influenced by the investigator's knowledge about facilities and human factors. Particularly, when it comes to finding commonalities among human error event data accumulated in an organization, reliable and useful results do not appear unless the analysis of individual human error event is based on the unified mindset. Hence, in 2006, we developed a framework (see Figure 6.2) for exploring causal factors of trigger actions (Hirotsu et al., 2006).

This framework was developed based on the model of accident investigation (stages in the development and investigation of an organizational accident) described by Reason (1997), as well as our experiences in event analysis. Each item of this framework was defined while considering work in a nuclear power plant. Moreover, the causal factors reference list (Figure 6.3) was summarized based on this framework, in order to assist investigators who do not have sufficient knowledge about human factors in identifying causal factors.

This framework is applied to a causal analysis (Step 3 above) after identifying trigger actions. First, one examines the factors concerning personnel involved at implementation phase, which is the working level. These factors concern workers or work group members. Next, one examines local workplace factors such as task demands and work environment. After that, one examines work control such as preparing procedures and work packages. Finally, one discusses management factors such as training, quality control, and safety culture. We see from Figure 6.2 that the scope of the factors discussed is gradually broadened. Thus, analyzing various factors, ranging from those directly related to errors to management, leads to identification of problems in the whole organization.

THE PROCEDURE OF HINT/J-HPES

In order to permit investigators who do not have sufficient knowledge concerning human factors and analysis experience to identify causal factors and to develop countermeasures, we reviewed the analytic procedure of the J-HPES (Takano, Sawayanagi, & Kabetani, 1994) by reflecting on the basic framework of Figure 6.2 (Hirotsu et al., 2006). The revised procedure, named HINT/J-HPES, comprises four stages. ("HINT" is not an acronym, but was added to the name of the method because the revised version includes enhanced hints, in the form of the basic framework, for causal analysis.) Stages 1 and 4 have not changed from those of the original J-HPES. Gathering information for Stage 2 has been enhanced by using the causal factor reference list (Figure 6.3), with the basic framework (Figure 6.2) as a reference. The framework (Figure 6.2) has also been applied to the causal analysis (Stage 3) to guide the search down to the management factors.

STAGE 1: UNDERSTANDING OF EVENTS

A timeline of what happened before the event is consolidated in the form of an event sequential table (see Figure 6.4). In the second column of this table, abnormal

Work Phase	Types	Categories	Subcategories (example)
Implementation phase	Human	Communication	Pre-job briefing (Work direction etc.) Communication during work (Reporting, communicating, counseling etc.) Turnover (Coordination, turnover etc.)
		Work practice	Execution (Procedure use, self checking, housekeeping etc.) Check (Hold point etc.) Cleanup (Cleanup of workplace etc.)
		Psychological	Memory (Memory lapse, preoccupation etc.) Emotion (Pressure, effort, impatience, self-
Management phase		Rules (Department/Plant)	Rules & regulation, guidelines, etc.

FIGURE 6.3 Causal factors reference list.

Date/ Time	Abnormal State	Work Phase	Acts & Communications					Problems	
			Planning manager	Job supervisor	Worker A	Worker B		Trigger Actions	Contributing Actions
5/14		Planning	Work permit	Prepare procedure					Compiled procedure based on examples of maintenance during power outage
5/18 10:30		Implementa-tion		Supervise other job	Check terminal number	Inspect controller A			Worker B used un-insulated screwdrivers
14:00				Supervise other job	Check terminal number	Inspect controller B			Worker B held two screwdrivers in one hand
14:30	Blown fuse					Survey controller B		Made the terminals short-circuit	

FIGURE 6.4   Event sequential table (with examples).

machinery states are described. Next, each abnormal state is examined to see whether it was due to a human activity. If so, the activity is defined as a trigger action. Then, a series of activities of the workers who were associated with the trigger action is described. Finally, actions or communications that either induced the trigger action or led to it being overlooked are specified as contributing actions.

STAGE 2: GATHERING AND CLASSIFYING INFORMATION ON CAUSAL FACTORS

In this stage, interviews and field investigations concerning trigger actions and contributing actions clarified in Stage 1 are carried out in reference to the causal factor categories listed on the form for causal factor data (Figure 6.5). This form is based on the causal factors reference list (Figure 6.3). Any collected information is classified and filled into the form. Contributing actions identified in Stage 1 are classified as causal factors of "communication" or "work practice" of "implementation phase [human]" or "planning, preparation, & evaluation phase [human]." After the gathered information is filled into the form, possible contributions to the trigger actions of each column are evaluated in discussions among investigators, and the result is recorded in the evaluation column. If the description seems to have contributed directly or indirectly to the occurrence of trigger actions, a Y will be placed in the evaluation column. If not, an N will be placed in the evaluation column. Neglecting the human factors viewpoint in data gathering is prevented by confirming that information corresponding to each category of this form is present.

STAGE 3: CAUSAL ANALYSIS

In this stage, the conceivable causal factors are analyzed to draw up a causal relation chart in the format shown in Figure 6.6. The basic idea of the analysis is to ask the question "why?" repeatedly and to use the basic framework of Figure 6.2 to deepen the analysis. For each trigger action, a fault tree-style causal relation chart is created to examine the causal factors listed in Stage 2. At this point in time, causal factors that form the basis of each trigger action are explored as far as possible in accordance with the five levels of the basic framework shown in Figure 6.2. The part where the examination is insufficient (dotted line frame in Figure 6.6) can be clarified by associating a classification framework to the causal relation chart.

The analysis result can be fulfilled by supplementing such shortages with additional data collections. Through this procedure, investigators will become aware of causal factors they did not take notice of in the previous J-HPES procedure and will thereby be able to obtain more satisfying results. In most cases, latent problems in the management phase are not mentioned, particularly during the interview in Stage 2. By referring to the basic framework, investigators try to find out the latent factors in management that may have contributed to the causal factors listed in Stage 2 and linked in the causal relation chart. An example of a causal relation chart can be found in Figure 6.7.

Work Phase	Types	Categories	Factors	Evaluation
Implementation Phase	Human	Communication	Worker A failed to warn worker B about improper handling of tools	Y
		Work Practice	Worker B held two screwdrivers in one hand	Y
			Worker B used an un-insulated screwdriver	Y
		Psychological Factors	Work group members were not made aware of short circuit danger during pre-job briefing	Y
		Physical, Physiological Factors	Work group members had no problem in their health	N
		Knowledge, Skill	Worker B had no experience in handling live circuits	Y
	Work	Work Characteristic	The task was monotonous and repetitive	Y
		Work Hour	The schedule was tight	Y
Planning Phase	Human	Communication	Supervisor had not told workers about short circuit danger	Y
Management Phase		Work Practice	Supervisor compiled the procedure based on examples	Y
		Organizational Culture		
		Rules (Department/Plant)		

FIGURE 6.5   Form for causal factor data (with examples).

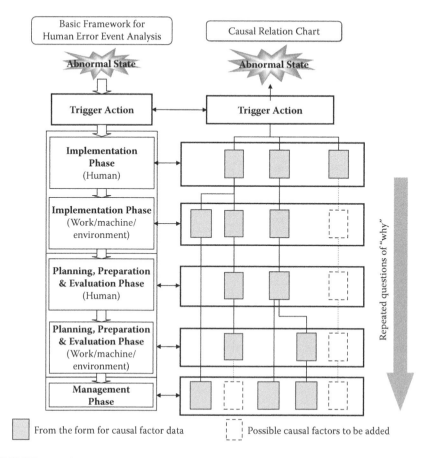

FIGURE 6.6 Causal relation chart (HINT/J-HPES).

STAGE 4: PROPOSING COUNTERMEASURES

The fourth and final phase in HINT/J-HPES is to develop countermeasures for the purpose of correcting problems identified in Stage 3. Table 6.1 presents a form for countermeasure proposal.

First, specific countermeasures are developed for the trigger action. The countermeasures are selected based on the following criteria:

- Providing a means of averting an abnormal state being caused by a trigger action
- Providing a means of preventing the occurrence of a trigger action

Next, specific countermeasures are developed for each causal factor identified in Stage 3. The countermeasures are selected based on the following criteria:

- Providing a means of averting a harmful effect of a causal factor
- Providing a means of preventing the occurrence of a causal factor

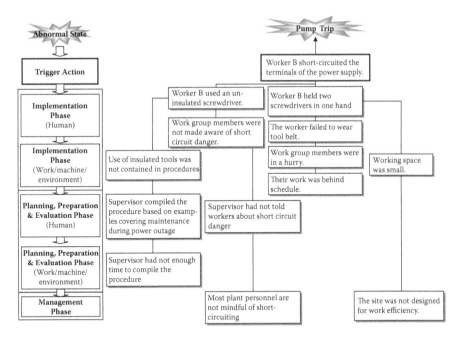

FIGURE 6.7   An example of causal relation chart (part).

TABLE 6.1
Form for Countermeasure Proposal

Level	Description of Countermeasures	
	Equipment Improvement	
	Working Environment Improvement	
	Improvement of Procedures and Work Management Method	
	Training/Familiarizing	
Level 1 [Trigger Action]	Equipment Improvement	
	Working Environment Improvement	
	Improvement of Procedures and Work Management Method	
Level 2 [Implementation Phase]	Training/Familiarizing	
Level 3 [Planning, Preparation and Evaluation Phase]	Equipment Improvement	
	Working Environment Improvement	
	Improvement of Procedures and Work Management Method	
	Training/Familiarizing	
Level 4 [Management Phase]		

Level	Description of Countermeasures	
Level 1 (Trigger Action)	Equipment Improvement	–
	Working Environment Improvement	(1) Have all bare wiring in control cubicles covered by insulation.
	Improvement of Procedures & Work Management Method	(2) Stipulate by rule to have this kind of job done with power cut out.
	Training/Familiarizing	–
Level 2 (Implementation Phase)	Equipment Improvement	–
	Working Environment Improvement	Same as (1) above in Level 1
	Improvement of Procedures & Work Management Method	Same as (2) above in Level 1 (3) Add provision in textbooks for preventing electrical accidents: Avoid work on live circuit in so far as possible. When indispensable, only use insulated tools. (4)
Level 4 (Management Phase)		(13) Include considerations of maintainability in the design of cubicles. Same as (3)(4)(9)(10) above in Level 2

FIGURE 6.8 An example of countermeasure proposal (incomplete).

In addition, the following four categories of countermeasures are shown on the form for countermeasure proposal, in the order of the effect of the prevention of recurrence:

1. Equipment improvement (improvement of the machinery, fail-safe, etc.)
2. Working environment improvement (indication bill, the improvement of the tool for operation, etc.)
3. Improvement of procedures and work management method (improvement of work management method, revising procedures, etc.)
4. Training/familiarizing (training for safety work, provision of information, calling for attention, etc.)

These categories can assist investigators in coming up with various countermeasures. An example of a countermeasure proposal can be found in Figure 6.8. It looks as if many countermeasures could be proposed by following this procedure. However, because each countermeasure can address several causal factors, the number of countermeasures will not be proportional to the number of causal factors. Countermeasures selected for implementation can be prioritized if the resources are limited.

LOOKING FOR COMMONALITIES AMONG HUMAN ERROR EVENTS

Because serious events are rare, it is essential to be able to recognize latent weakness by looking for commonalities among minor human error events (i.e., events where the outcomes are not significant), and from this to address the problems with

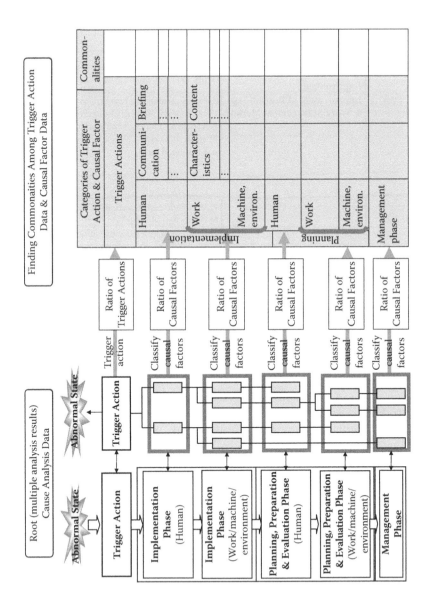

FIGURE 6.9 An image of finding commonalities among event databased on the basic framework.

the entire organization. We therefore developed a procedure for finding commonalities among accumulated analysis results of human error events (Hirotsu et al., 2006). Using the basic framework introduced in Figure 6.2, we created a diagram that shows the procedure for finding commonalities among events (Figure 6.9).

Regarding trigger actions, trigger action data of analysis results of multiple events within a certain time frame (e.g., in the past year) are classified according to the trigger action categories (Figure 6.10), which were obtained by classifying trigger action data of human error events at Japanese nuclear power plants. The most prevalent trigger action categories are determined on the basis of the ratio of each category occupying the total number of error occurrences. This means that if three of ten trigger actions are classified in a category called "skip of an operation step," then the category occupies 30% of the total trigger actions; this should be recognized as a characteristic error type.

As to the causal factors, the data from the analysis results of multiple human error events can be classified according to the subcategories of causal factors (Figure 6.3). The most prevalent causal factor subcategories are determined on the basis of frequency of occurrence of each subcategory against the total number of error occurrences. In other words, the finding that the factors concerning "communication during work" are related to seven of ten trigger actions should be interpreted to mean that "it is a characteristic of the organization that is common to 70% of the total number of trigger actions there." By this method, we can avoid the analysis result being affected by just a few events with many causal factors involved.

Moreover, in terms of the characteristic trigger action categories and causal factor subcategories identified above, commonalities are extracted on the basis of

Trigger Action Mode of J-HPES	Categories for Operations	Categories for Maintenance
Omission	• Overlooking abnormal condition • Skip of an operation step	• Overlooking abnormal condition • Skip of a maintenance step • Omission of preventive maintenance and monitoring
Drop, contingence, falling, intrusion	• Hit/knock together	• Hit/knock together • Contacting bare live part • Fall • Foreign material intrusion
Wrong object	• Wrong unit/train/component	• Misconnetion and miswiring of terminal • Wrong unit/train/component
Improper manipulation/ work amount	• Insufficient manipulation of valve	• Insufficient tightening of terminals • Insufficient torque of bolts

FIGURE 6.10   Categories of trigger action (for nuclear power plant).

description of each concerning action and causal factor. Take the case of "communication during work," where there are possibilities such as inadequate reporting and the vague instructions of the boss. We therefore further examine the descriptions of causal factors related to seven trigger actions classified as "communication during work" in order to extract concretely the commonalities of inappropriate communication.

This method is effective because the trigger actions and causal factors affecting multiple events will be recognized as problems, even if they are perceived as having little significance for an individual event. For example, a causal factor such as "using inappropriate tool" of a minor event might be addressed by the workgroup performing similar work activities. However, if the recurrence of similar problems (factors) is identified by analyzing multiple events, the basis for managing tools or instruction for selecting them will be fundamentally reviewed. Thus, identifying problems as the characteristics of an organization and actively coping with them will improve the resilience of the entire organization against human errors.

MANAGEMENT PROCESS FOR HUMAN ERROR PREVENTION

From the human factors perspective, the problems in an organization change as time goes by. Because of this, it is necessary for an organization to try continuously to grasp latent problems and to set up activities to address them appropriately. One way of doing this is by means of error management (see, for example, Reason & Hobbs, 2003; CCPS, 1994), which is comprised of proactive and preventive activities for error prevention. In order to conduct these activities systematically for continuous improvement, a plan–do–check–act (PDCA) management approach is also useful. On the basis of these ideas, we developed an error management process that utilizes human error event data (see Figure 6.11; Hirotsu et al., 2006). Event analysis (analysis

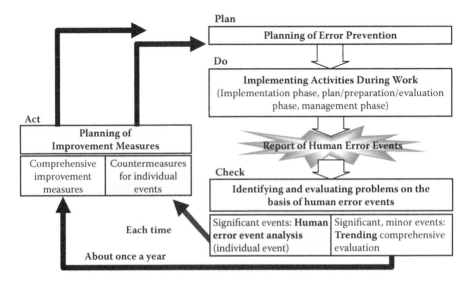

FIGURE 6.11   Error management process that utilizes human error event data.

of individual human error events and the analysis for finding commonalities among the multiple analysis results) is placed in the *check* stage of this process.

The following sections summarize the procedure for each stage of the PDCA approach in error management.

PLAN: PLAN ERROR PREVENTION ACTIVITIES

The *plan* phase is the first major step in the error management process. The focus is on how to identify problems in an organization and develop plans for error prevention activities to address them.

Because there is no silver bullet for the prevention of human errors, it is important to conduct activities, steadily and continuously, that are based on the standpoints of quality management and safety management. In order to prevent these basic activities from becoming habits, management should revise priority activities every year. Priority activities should be aimed at improving the basic activities and addressing short-term problems from the human factors perspective. Problems in an organization that have been clarified by analysis of in-house events or questionnaire surveys can serve as the clue for choosing priority activities. If activities that can solve the problems are found among the basic activities, those should be chosen as priority activities, and detailed plans for conducting them should be developed. If that is not the case, a new activity could be introduced with reference to best practices of other organizations. Management should decide on a policy for the priority activities and each department should define the plans for conducting the activities in detail on the basis of characteristics of their work. Management should collect the priority activity plans of each department and review measures to support them.

DO: IMPLEMENT ACTIVITIES DURING WORK

The second step in the management process is work execution or implementation. In this phase, the basic activities and priority activities identified in the *plan* phase are implemented. When an event triggered by a human error occurs in the course of daily duties, personnel report it to the organization. The event information, including minor events, becomes the base for facilitating this process. Table 6.2 shows an example of a human error event reporting sheet. This form is supposed to be used by the personnel involved, to report events to the organization. By using this form, the personnel can specify trigger actions and causal factors of the implementation phase and can also develop the countermeasures for them. This sheet can also be used to spread the event information throughout the organization. To improve the reliability of the data collection using this form, the causal factor reference list (Figure 6.3) should be referenced, and the filled-out form should be checked by personnel who are familiar with human factors and event analysis.

CHECK: IDENTIFY AND EVALUATE PROBLEMS ON THE BASIS OF HUMAN ERROR EVENTS

The third phase is the identification and evaluation (*check*) of problems from the perspective of human factors. In this phase, human error event data are utilized

TABLE 6.2
Human Error Event Reporting Sheet

Title	
Trigger Action	
Event Description	
Figure /Photo	

Causal Factors		Countermeasures
Human		
Work		
Machine, environment		
Management		

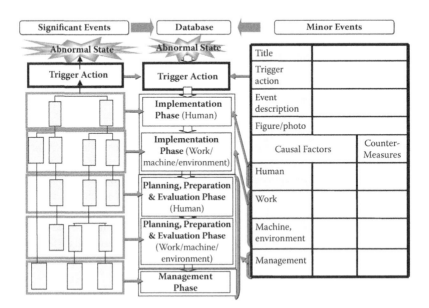

FIGURE 6.12 Accumulating event data.

directly by the organization in which the culture for reporting minor human factor events has been established. The investigation ideally should look at the fundamental causal factors, such as those concerning management, for all events reported with the human error event reporting sheet. However, when the number of events increases, it becomes a great burden to conduct a thorough analysis, so it is normally realistic to conduct the analysis only for significant events. Examples of significant events could be events with significant outcomes (reportable outcomes) or events that to a considerable extent involve problems in work control or management (such as latent errors).

If an event reported in the form of Table 6.2 is recognized as a significant human error event, an extensive analysis in accordance with HINT/J-HPES should be conducted in order to propose countermeasures for preventing recurrence. When this has been done, management should review the error prevention plan immediately. If extensive analysis cannot be done, data concerning trigger action and causal factors should be accumulated together with that of significant events (Figure 6.12), and the commonalities should then be extracted about once a year.

Then, improvements of problems identified in the *plan* phase should be checked by regular comprehensive evaluation. The extracted commonalities should be compared with the problems they were intended to solve with the priority activities identified in the *plan* phase and implemented in the *do* phase. If the same problems are not specified, it can be seen as evidence that the priority activities have been effective. Otherwise, there may be problems in identifying, planning, or conducting the priority activities, and the cause of these shall be identified.

ACT: PLAN IMPROVEMENT MEASURES

The fourth and final phase in the management process is the *act* phase, which is to plan comprehensive improvement measures based on the results of the *check* phase. As to the problems newly specified in the *check* phase, new priority activities are developed. When the same problems are specified repeatedly, it is necessary to develop a strategy for improving ineffective priority activities, based on the reason why the activity plan has not worked. The result of this examination will be reflected in the *plan* phase of the next cycle. In addition, management will examine the necessity of the additional activity from error prevention activities trend of error prevention activities inside and outside the industry and will finally summarize a plan as an organization.

TO ESTABLISH AN EFFECTIVE PROCESS FOR REDUCING ERRORS

In order to create safety culture and to become resilient to human errors, workers as well as all other personnel levels, including management, need to take part in the activities. Continuous improvements are attained when an organization unifies the direction of multiple departments and personnel levels and evaluates them to make sure they operate systematically. Therefore, as suggested by the process presented here, the development of activities that are aligned to the strategy of error prevention as an organization is considered to be effective. This strategy should provide guidelines that are more definite than some vague advice, such as minimizing the number of error occurrence, and should aim at overcoming weaknesses that trigger problems.

To establish an effective process for error prevention, it is not enough to examine methods only for data analysis, evaluation and utilization. As preconditions for these activities, a structure and culture for reporting high quality event data is essential. In other words, the starting point for managers in conducting these activities is to understand the importance of reporting events and to establish rules and programs for event data utilization. In addition, investigators have to master the analysis methods to some extent and should also be able to analyze multiple events with a coherent mindset from the human factors point of view. This cannot be achieved by conducting analysis as a side job. For this reason, it is necessary to have personnel who can devote themselves to these activities.

To motivate an organization with an established culture for reporting events, including minor ones, it is convincing to show the problems identified based on these reports. In order further to identify problems at the *plan* and *check* phases, various methods are available, such as field observations of workers' behaviors, questionnaires, and interviews concerning safety consciousness of individuals. For the future, it is clearly desirable to evaluate problems comprehensively, including results from other evaluation methods.

CONCLUSION

A basic framework for human error event analysis has been developed and incorporated into the J-HPES and into analyses for finding commonalities among human error events. As a result, these analysis methods support the investigators to clarify latent problems in the organization, such as work control process and inadequate practices of workers. Furthermore, the analysis based on the basic framework also enables the development of a management process that utilizes human error event data.

The proposal of HINT/J-HPES and the use of the analysis results described previously have been developed on the basis of examples at domestic and foreign nuclear power plants. But it should also be possible to apply these methods to other industries by making the necessary modifications. Further work should be carried out on how to adapt them to other domains such as electric power transmission, power system operation, and distribution of electric power. As to the error management process, incorporation of additional evaluation methods, such as behavior observation and questionnaires to enhance the *check* stage, is under study.

High reliability organizations make an effort to create a reporting culture and work hard to extract the most value from what little data they have (Reason & Hobbs, 2003). It can be argued that the systematic activities of analyzing and utilizing event data proposed here are effective means to improve the safety consciousness of each individual, as well as effective to maintain and improve the safety and reliability of facilities.

REFERENCES

Barnes, V. & Haagensen, B. (2001). *The human performance evaluation process: A resource for reviewing the identification and resolution of human performance problems.* NUREG/CR-6751. Washington, DC: U.S. Nuclear Regulatory Commission.

Center for Chemical Process Safety of the American Institute of Chemical Engineers (CCPS) (1994). *Guidelines for preventing human error in process safety.* New York: American Institute of Chemical Engineers.

Hirotsu, Y., Ebisu, M., Aikawa, T. & Matsubara, K. (2006). *A trend analysis of human error events for proactive prevention of accidents — Methodology development and effective utilization,* CRIEPI Rep. Y05010 (in Japanese).

Paul-Stüve, T. (2005). *A practical guide to the why–because analysis method — Performing a why–because analysis.* (http://www.rvs.uni-bielefeld.de/research/WBA/WBA-Guide.pdf, accessed August 11, 2008).

Reason, J. (1997). Managing the risks of organizational accidents. Aldershot, UK: Ashgate.

Reason, J. & Hobbs, A. (2003). *Managing maintenance error: A practical guide.* Aldershot, UK: Ashgate.

Takano, K., Sawayanagi, K. & Kabetani. T. (1994). System for analyzing and evaluating human-related nuclear power plant incidents. *Journal of Nuclear Science Technology,* 31, 894–913.

7 Beyond Procedures
Development and Use of the SAFER Method

Yutaka Furuhama

CONTENTS

THE TEPCO NOTION OF HUMAN FACTORS ENGINEERING

In the present, owing to extensive activities in ergonomics/human factors over a half century, and at the sacrifice of many tragic accidents, the general idea of "to err is human" seems widely recognized in our society. Facing our reality, however, it seems inappropriate to say that incident/accident analysis and corrective actions are well-performed based on this knowledge. Despite improper circumstances that are not compatible with human characteristics, or even exceed human capability, blame for the accident, in many cases, is finally put on the persons concerned. Under the legal notion of "professional negligence" in Japan, the persons concerned are often arrested and prosecuted. Examples of this attribution include a fatal medical accident in Kyoto in which ethanol was mistakenly injected into a respirator (*The Japan Times*, 2000). In another incident, a slip of the tongue by flight control resulted in a severe near-miss accident in the skies over Shizuoka in 2001 (Aviation Safety Network, n.d.). And, additionally, a fatal railroad crossing accident where the toll bars were manually opened happened in Tokyo in 2005 (Cabinet Office, 2005). Such views of accidents that attribute causes to individuals tends to lead to countermeasures, depending on one's mental prowess or courage, falling into a vicious circle of higher demands on individuals and a higher probability of erroneous action.

How can we break this vicious circle? Different from an intentional crime, an error and its consequences is never desired by the person concerned. Thus, it is natural to think that some factors induced him or her to commit the error. Investigation of those factors and to subsequently take measures to prevent them should, therefore, be the object of an incident/accident analysis method that aims to prevent the recurrence of undesirable results.

The image in Figure 7.1 is often used to illustrate some of our cognitive characteristics. In this figure, the symbols in the upper row can be read as the letters A B C, whereas the symbols in the lower row look more like the numbers 12 13 14. But the middle symbol is the same in both rows! Yet we see it as the letter B in the upper row and as the number 13 in the lower row. This is due to the neighboring symbols, and the example illustrates that our recognition and interpretation can be strongly affected by the surroundings. Because of such characteristics, we can sometimes quickly recognize various situations without much effort. On the other hand, if the surroundings induce a misunderstanding, we can sometimes also wrongly recognize something.

Imagine that a map of the location of fire exits in a hotel is found on the inside of the door of your room, No. 212 (see Figure 7.2). In case of an emergency, can you correctly find out from, a glance which way to go? Open the door and then turn right — or turn left — to find the nearest stairs? For the even-numbered rooms, right and left as shown on the sign are opposite of the real direction when one opens the door to the corridor, and it is therefore likely that you will go the wrong way. In this case we can

FIGURE 7.1   How could you read these symbols.

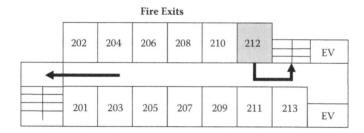

FIGURE 7.2   A misleading map (imagine being found inside the door of room 212).

say that an erroneous action (namely, going in an undesirable direction) is induced by the map, which is oriented in conflict with the principle of "natural mapping."

The careful consideration of several accidents has revealed many cases of such "error-inducing factors." On the basis of these findings, we propose that human errors are induced when the surrounding factors are incompatible with the physical, physiological, cognitive, and psychological characteristics of the human nature. In other words, human error is not a cause, but is a result or consequence of these inducing factors. This is the basic meaning of what the Tokyo Electric Power Company's (TEPCO) Human Factors Group (HFG) refers to as *human factors engineering*. This corresponds to Lewin's equation, which says that human behavior is produced by the interaction between humans and surroundings (Sansone, Morf, & Panter, 2004).

In 1994, we proposed the m-SHEL model (Figure 7.3) as a conceptual model for this view on human factors engineering. The m-SHEL model is based on the SHEL model, developed by Hawkins (1987). The letter L (liveware) in the center represents a person carrying out some actions, and the other letters denote the various surrounding factors: hardware (H), software (S), other liveware (L), environment (E), and management (m).

The representation of this model in Figure 7.3 has two important features: (1) the L is located in the center, and (2) every letter is enclosed by an uneven or wiggly square. The position of the L reflects the human-centered concept, and the uneven or wiggly squares mean that each factor has its own characteristics; for example, human–machine interface design for hardware (H), writing style for software (S),

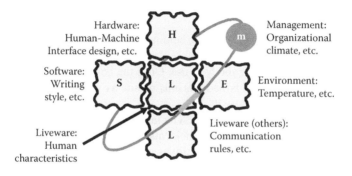

FIGURE 7.3   m-SHEL model (designed by R. Kwano).

communication rules for other liveware (L), and so forth. The gaps between the squares represent error-inducing conditions. According to the model, there are two ways to bridge them. The first is to adapt the L in the center to the surroundings, for instance, by training persons to accommodate to circumstances. The second is the opposite, which represents the so-called human-centered system design; in other words, adapt the surroundings to the center. Although many people think that the first solution is good enough, this way of bridging the gap is neither a universal nor a permanent solution. The reason for this is that humans by nature are susceptible to the surroundings, so that even a skilled person may commit an error if the surrounding, error-inducing factors are strong enough. The basic policy of human factors engineering should therefore be to make the surrounding factors compatible to human characteristics, in order not to induce undesirable action and consequences.

OUTLINE OF THE INCIDENT/ACCIDENT ANALYSIS METHOD — SAFER

This section presents an outline of SAFER, an incident/accident analysis method based on the human factors engineering principles described previously. The name SAFER, meaning Systematic Approach For Error Reduction, was originally developed in 1997 as H^2-SAFER by the TEPCO HFG (see Yoshizawa, 1999). (H^2 stands for Hiyari-Hatto, which is the Japanese term for near misses.) It was considerably improved in 2003 and renamed SAFER. Refinements are still continuing in an effort to provide a better and more usable method.

BACKGROUND AND FEATURES OF SAFER

H^2-SAFER was a product of field-oriented or fact-oriented thinking. Because a number of error-inducing factors were found by the analysis of real accidents, it led to the idea that analyzing an accident should reveal the whole set of background factors. Effective analysis and corrective activities are usually performed by people on the site rather than by external method specialists. There is therefore a need for a handy analysis method that is easy for the on-site people to use and that helps them reveal the whole set of background factors for an incident or accident. This motivation for developing H^2-SAFER remained unchanged during the revision that changed it to SAFER.

The main features of SAFER can be summarized by the following points:

- It is convenient for everybody to use: Once the basic notions and the steps of analysis have been learned, persons on-site as well as the specialists in methodology can use it easily.
- It is applicable to various events: The target events cover everything from serious accidents to near-miss incidents and are not restricted to human errors but include problems in facilities and organizational matters.
- It is useful for developing a common way of thinking: The basic notion of human factors engineering, and the knowledge about how surroundings

can lead to accidents, is far more important than procedures and formats. The use of SAFER can help a person acquire the underlying notion, the viewpoints, and the way of thinking, and to share them on-site or in the office.

THREE STAGES IN THE FRAMEWORK OF SAFER

Neither a good analysis of accidents nor effective means of correction can be produced if one looks only at the moment when the erroneous actions and/or accidents happened. What happened, the errors or accidents, is the result of something else. In analyzing an event we therefore have to trace the history of how a trigger was induced and how it developed into the final consequences. We have to reveal the whole set of background factors related to the history, and this should be the grounds for effective countermeasures. To realize this idea, we proposed the following three stages, which constitute the framework of SAFER:

1. Fact-finding: Develop the right understanding of what happened during the event and find the related facts.

    A first step is to arrange information and make an event flow chart, both to correctly understand the details of the event and to share them among participants. One way is to align the persons and facilities concerned along the horizontal axis of the chart and to show the flow of time on the vertical axis. The next step is to enter every piece of information onto the chart using simple phrases, and then to connect each piece with arrows to clearly show the flow and development of the event.

2. Logical investigation: Use multi-sided analyses to find the causality among the various background factors behind the event.

    Based on the information in the chart, one should logically investigate the background factors behind the event in order to produce a background factors causality diagram, which represents the causal relations among the factors. This diagram is similar to the fault tree that is generally used in describing the failure analysis of mechanical systems. Based on the notion of human factors engineering, it provides a complete view of the background factors and shows how various factors are linked or interrelated and how they finally resulted in the event. It is necessary to make a proper diagram (i.e., to represent all factors and their causal relations as correctly as possible) in order to develop effective countermeasures. On the basis of our on-site experience, we have therefore prepared considerable guidance, some of which will be mentioned in the following sections.

3. Preventive measures against background factors: Consider how to cut off the causality in order to prevent the event from recurring.

    As the last step, try to develop preventive measures to cut off the causality among the background factors that caused the event, according to the background factors in the diagram. Then decide on the order of priority to implement preventive measures based on the evaluation of their effect, residual risk, and difficulty of execution, such as cost and lead time. A proper diagram logically shows the candidate factors that can be used to take preventive measures. Together with the evaluation of their effect and residual risk, this provides a comprehensive viewpoint that enables us to decide efficiently on useful preventive measures. We have also at this stage prepared considerable guidance to serve as a help to think about effective preventive measures, as well as a method to evaluate them. All of this is based on the notion of human factors engineering.

The importance of these three stages derives from the purpose of analysis, which is to prevent the undesirable event from recurring. In order to prevent an event from recurring, countermeasures are needed to cut off the causality relations among the background factors that induced the event. This demands a comprehensive and correct representation of the factors and causality relations, which in turn requires grasping a wide range of facts related to the event.

The SAFER procedure embodies these three stages and further splits them into eight steps, as shown in Figure 7.4. The first step, understanding the notion of human factors engineering, occurs before the first stage. The first stage corresponds to Step 2, the second stage to Steps 3 and 4, and the third stage to Steps 5 and 6, respectively. Steps 7 and 8 lie beyond the SAFER desktop analysis and are not easy to generalize as method; therefore, we provide only a few remarks for their implementation. Note that in the actual incident/accident analysis, Steps 2–4 (or sometimes Steps 2–6) are not necessarily sequential but may be repetitive, because not all information about the event is ready in advance.

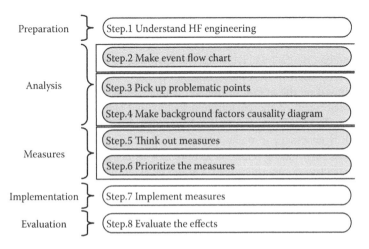

FIGURE 7.4 Procedure of SAFER.

IMPROVEMENT OF SAFER BASED ON EXPERIENCES ON-SITE

In order to contribute to reducing incidents and accidents related to human erroneous action, the Human Factors Group has promoted SAFER within TEPCO. Through this activity we have found many cases where background factors were not properly investigated and/or where preventive measures seemed ineffective, even though the persons doing the analysis followed the SAFER procedure. The following briefly describes some typical issues that were found, together with the corresponding improvements that have been made to SAFER during the last few years:

Issue 1. Place more emphasis on procedure and model and less consideration on why errors or accidents were induced.

This issue relates to Steps 2–6 in the SAFER procedure represented in Figure 7.4. One typical misuse of SAFER is that background factors are classified by the m-SHEL model. Although the use of the m-SHEL model or the classification of background factors is not bad in itself, neither of them is essential for using SAFER. It is more important to make a logical investigation of the background factors that induced the errors or accidents than to classify them. The m-SHEL model is furthermore not a strict model that can be used to prescribe viewpoints but more like a loose framework of reference that can be used to bring out multiple viewpoints in investigation.

Preventing an over-adherence to procedure and a misuse of the model is a difficult issue in performing an event analysis, and a simple improvement of the procedure or the guidance will not be sufficient. We therefore first tried to improve the procedure by adding as a first step the idea of understanding the notion of human factors engineering (Figure 7.4). Because the problem was an outcome of persistently following the procedure, we explicitly built in the basic notion as the first step. Besides this improvement, we prepared some guidance for how to make a background factors causality diagram, such as "Do not stop searching for background factors when you find a factor related to a person's action or consciousness." This continuation rule encourages the analyst to pay more attention to surroundings and to the context that induced the consequences.

Issue 2. Unclear causality between consequences and background factors, or poor grounds to show the effectiveness of preventive measures.

This issue mainly relates to Steps 4 and 5 in the SAFER procedure (Figure 7.4). An example of unclear causality is, for instance, to ascribe an outcome (e.g., a worker received a burn) to an arbitrary operation of the injured worker. Although this sometimes might be a reasonable guess, it does not explain why the arbitrary operation led to the outcome (the burn). The uncertainty of such a cause–effect relation weakens the basis for claiming that countermeasures taken to prevent the operation in question will be effective to prevent future instances of burns.

This issue of unclear causality has long been recognized as important, and we have used our experience to prepare advice on how to appropriately

Cause & Effect Diagram versus Background Factors Causality Diagram

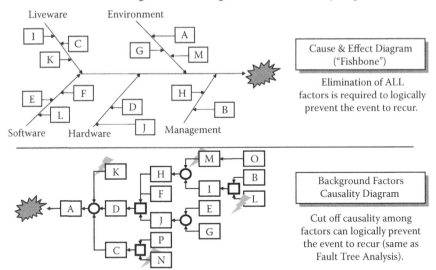

FIGURE 7.5   An example of illustrative material to explain the basic principle of SAFER.

investigate causality, as mentioned in the following sections. This underlines that the basic principle of SAFER is to reveal the causality among background factors, and to produce preventive measures to cut off the causality. Although this in some ways is similar to a fault tree analysis, as mentioned previously, the uncertainty of human actions means that it does not require the same detailed and strict procedure. In order to explain this principle in an easily understandable way, we prepared some illustrative materials, as shown in Figure 7.5. This shows the difference between our background factors causality diagram and a cause-and-effect diagram that often is used in quality control activities.

Issue 3. Logical jumps in investigating the causality between background factors.

This issue relates to Step 4 in the SAFER procedure. It is important that the logical investigation of background factors (i.e., an investigation of the underlying causality) is free of logical jumps, because otherwise the effectiveness of a proposed measure to cut off the causality cannot be justified. Basically, the investigation is performed by asking why this result or factor is induced, but it is difficult to think in this way without sometimes making logical jumps. We therefore prepared some guiding principles on how to support a logical investigation, as illustrated in Figure 7.6. For instance, subdivide what happened and observe it physically or verify the logic of causality by backtrack; in other words, if it is found that A is caused by B, then verify whether B really causes A. By using the guiding principles, the determination of a direct cause of, for example, a burn accident should not point to

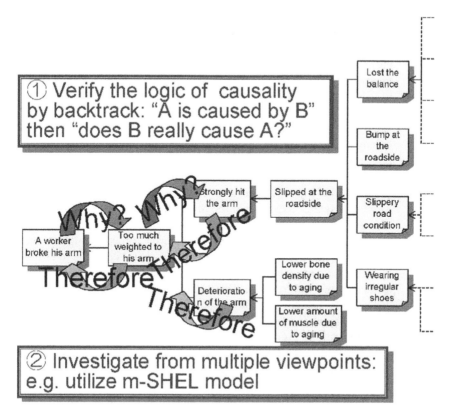

FIGURE 7.6 An example of know-how to make a proper causality diagram.

the worker's arbitrary operation, as mentioned previously, but instead point to the coexistence of "something being hot" and "that something is touched." This is a more appropriate explanation of how the burn physically occurred. It also verifies the causality, because touching something hot is certain to result in or cause a burn, whereas an arbitrary operation is not.

Issue 4. Many requests for a comprehensive guide on how to think out preventive measures.

This issue relates to Steps 4 and 5 in the SAFER procedure. A guideline named H^2-GUIDE was given in H^2-SAFER. This guideline covered a wide range of ideas for preventive measures, from "elimination" to "preparation," based on the notion of an error-proof technique (i.e., trying to make sure something is error proof). Although the H^2-GUIDE was still very useful, we improved it to make it more comprehensive. It is now an easy-to-use guide that consists of eleven steps, as shown in Figure 7.7. At the same time, we also renamed it simply "GUIDE." The improvements came about in the following way. We first specified that the object of countermeasures is to prevent or minimize damage resulting from accidents related to human erroneous action. We then introduced a distinction between two phases, prevention of errors and mitigation of effects, and two approaches, improvement of surrounding factors and improvement of individual abilities (individualistic countermeasures). This altogether resulted in the eleven steps shown in Figure 7.7. A detailed explanation of this solution is provided in Kawano (2006). (The individualistic countermeasures do not refer to individual psychological issues, but rather to established human factors principles and cognitive models of human behavior.)

Issue 5. Inadequate examination of candidate preventive measures may lead to a preference for individualistic countermeasures.

This issue relates to Step 6 in the SAFER procedure. When the candidate preventive measures are evaluated, individualistic countermeasures are often chosen even when countermeasures to improve the surroundings are present. This may be because it is common to examine preventive measures using their cost rather than their effectiveness as a criterion. One reason is that the true effectiveness of a measure generally is difficult to evaluate quantitatively. Another reason is that the thinking often seems to focus on the binary choice between taking some measure to prevent a recurrence and taking no measure.

In order to overcome this, we proposed a quasi-quantitative evaluation of the effectiveness of a measure, with GUIDE (Figure 7.7) as a basis, grading each on a scale from one to ten points. Residual risks and side effects of a measure are also introduced in order to promote a risk-oriented evaluation, for instance, to consider how much a measure decreases the risk of occurrence of an accident and/or damage. A combination of these

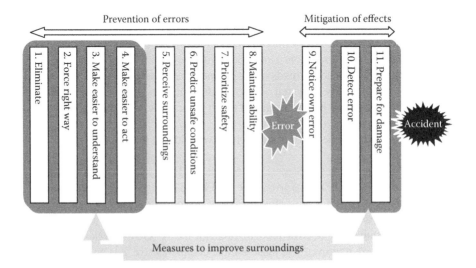

FIGURE 7.7   A guideline to think out preventive measures.

evaluations, with the difficulty of execution, such as cost and lead time, enables a realistic examination of countermeasures, while putting stress on their effect.

DESCRIPTION AND USAGE OF SAFER

Besides the improvements described above, we continuously evaluate and refine our experiences in a detailed manner and continue to develop instructional materials to improve SAFER, while keeping the basic notion and the overall framework unchanged. The following sections briefly describe the eight steps in the SAFER procedure and its usage as of July 2008.

STEP 1: UNDERSTAND HUMAN FACTORS ENGINEERING

The first step of SAFER is to understand the notion of human factors engineering, because this is the very basis for the other seven steps. This step is actually not an analysis activity as such but a preparation phase, and usually it is provided by an off-the-job course or lecture. We have developed some instructional materials, such as "e-learning contents," a reference book (Kawano, 2006), and presentation sheets. A standard set contains approximately sixty presentation sheets and is used for a one-hour course. This standard course consists of three parts: (1) providing arguments against the conventional view of human errors, (2) illustration of human characteristics and surrounding factors that affect human behavior, and (3) explanation of the notion of human factors engineering. In all three parts we include many small exercises and refer to many real incident/accident cases. Our experience has shown us that in order for persons to gain a clear understanding of human factors engineering and to utilize it as the basis of analysis activities, they must also be exposed to practice and case studies in addition to general knowledge. Our original "counting up game," which is a simple mental calculation in a context that induces the person to forget a figure carried, is an exercise in which people can experience their susceptibility to the context. Similarly, Figures 7.1 and 7.2 are examples that illustrate the effects of error-inducing surrounding factors on our cognitive characteristics.

STEP 2: MAKE EVENT FLOW CHART

This step is actually the first step of the event analysis work. The aim of this step is to understand properly what happened in the event and to share the information among participants. The process by which to make the event flow chart is very simple, as described in a previous section: line up the persons, facilities, etc., on the horizontal axis and show the flow of time on the vertical axis. After that, enter all pieces of information (actions, events) in the chart and draw arrows among them to show the information flow and the development of the overall event. It is useful to combine different sources of information to make the chart, such as evidence from an inspection of the scene, record of interviews with the persons concerned, documents about the task where the event occurred, etc. However, it is not necessary to have the complete information in advance, because the flexibility of the chart makes it possible to make additions and changes later on.

The experience of how to perform a good analysis is also included in this simple step. Some of this experience may appear very common or even trivial, but we have found that it is worthwhile to provide such guidance explicitly:

- Information should be traced into the past to a certain depth so that potential background factors can be considered. Examples are planning of tasks, alterations of design, and change of team members. It is important that the analysis is not limited to the scene of the event.
- It is recommended to include all types of information, supplementary explanations, and even presumptions in the chart, although a strict discrimination should be made between facts and other kinds of information.
- Each piece of information should be written briefly, possibly by using a simple phrase. This is good not only for the easy understanding and sharing of information, but also for maintaining a neutral attitude toward the facts.

In most cases, this step is performed using many tags on big sheets of paper, as shown in Figure 7.8. We have also developed a simple support tool for this step using Microsoft Excel.

FIGURE 7.8   An example of an event flow chart.

STEP 3: PICK UP PROBLEMATIC POINTS

Before beginning the investigation of background factors in order to make the causality diagram, one should pick up all possible problematic points from the event flow chart. This step is useful to make a thorough extraction of problems from the beginning to the end of the event. Some of these problems might not be obvious from the background factors found for the final consequences.

The possible problematic points not only include human actions but also deviations, unusual occurrences, and circumstances that might not be bad or problematic in themselves. As part of the work, it is possible to select pieces of information from the chart and transcribe them to other tags. For the convenience of the following analysis phase, it is recommended to add the subject of an action or a condition in this transcription. Each piece of information in the event flow chart need not refer to a different subject; several pieces in the same column may refer to the same subject.

STEP 4: MAKE A BACKGROUND FACTORS CAUSALITY DIAGRAM

The aim of this step is to provide an overall view or set of background factors that logically shows how a combination of these factors can lead to the event. The first step is to select one problematic point as the target for which a recurrence should be prevented. This is usually the final consequences, such as the damage to a facility or the violation of a regulation. Other problematic points, such as an unusual triggering action, can also be the target, depending on the purpose of analysis. The second step is to investigate the background factors in order to make a causality diagram based on the information in the event flow chart, using the target as the starting point. Most of the problematic points picked up in Step 3 are generally incorporated into the diagram as background factors. If some points remain, it means that heterogeneous problems have been left untouched; these may possibly become other targets to be the subject of new causality diagrams.

It is necessary to make a proper background factors causality diagram in order to consider effective preventive measures, and this is therefore an essential step in the SAFER analysis. The key issues in the investigation are logical thinking and the use of multiple viewpoints based on human factors engineering. As mentioned in the previous section, considerable experience is brought to bear on these issues, such as the verification of logic by backtracking causality, physical observations of what happened, continuation rules to prevent a premature stop of the analysis, and a reference to the m-SHEL model. Besides such basic guidance, we have also prepared advanced guidance, illustrated as follows:

- *Consideration of logical gates (i.e., AND and OR combinations of background factors).* If, for instance, there is an AND gate between factors that lead to an event, it is not necessary to develop countermeasures for every factor in order to prevent the event. This will in turn lead to an effective reduction in the number of preventive measures.

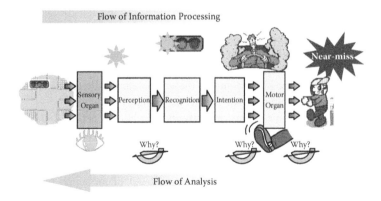

FIGURE 7.9   A reference to information processing model of human.

- *The use of a reference human information processing model to investigate background factors behind a person's actions.* Generally an action should be the product of a certain intention that depends on the person's knowledge, attention, recognition, and perception of information, all of which interact with the circumstances and the context (see Figure 7.9).
- *Investigation using multiple perspectives and positions, not only for the persons concerned but also for partners, witnesses, and victims of the event.* Although background factors for these other persons often are not considered, because they are beyond the range of preventive measures, they often provide good hints for what we should do as effective preventive measures.

In most cases, this step is also performed using many tags and large sheets of paper, as shown in Figure 7.10. The support tool mentioned previously is also available here.

STEP.5: THINK OUT PREVENTIVE MEASURES

After describing the overall set of background factors with the causality diagram, the next step is to think out preventive measures by which to cut off the causal relations that lead to the event. A typical misunderstanding is that preventive measures should be found for all factors at the very end of each chain of causality; namely, to all the factors shown at the right side of Figure 7.10. This solution certainly cuts off every chain leading to the event, so it is not entirely wrong. Yet it is neither essential nor efficient. It is instead important to cut off the chains anywhere possible in the diagram. If it is possible, for instance, to take concrete and effective countermeasures against a background factor close to the event (i.e., on the left side in Figure 7.10), they will be both more efficient and more effective. It may often be difficult to take such countermeasures because background factors near the event (on the left side of Figure 7.10) are consequences

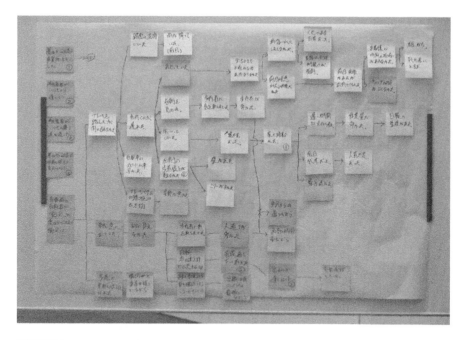

FIGURE 7.10   An example of background factors causality diagram.

rather than causes. In performing this step, it is important to show explicitly the correspondence between each measure and background factor, in order to clarify the aim of each measure.

Flexible and diverse ideas for preventive measures that differ from conventional examples or immediate restrictions are important at this step. To find these, we recommend the style of brainstorming, in order to make good use of other persons' ideas without criticizing them. In brainstorming, even wild ideas are welcome. Our comprehensive GUIDE (see also Figure 7.6) can help one think out effective and diverse countermeasures. We have also prepared a set of instructional materials with many examples, such as the one illustrated in Figure 7.11. Based on the notion of human factors engineering, the improvement of surrounding factors should precede the improvement of individual abilities; elimination-oriented ideas are also recommended as effective and reliable preventive measures. This view of priority is the basis for the evaluation of effectiveness, as described in the next step.

STEP 6: PRIORITIZE THE COUNTERMEASURES

Each measure proposed in Step 5 is prioritized by evaluating its effect, residual risk, side effect, and difficulty in execution. The effectiveness of a measure is graded on a scale from one to ten points according to the classification used in GUIDE (see Table 7.1). The numerical value of a point has no strict meaning, because this scale is mainly a numerical expression of the notions of GUIDE. These notions are that preventive measures are more effective if they depend less

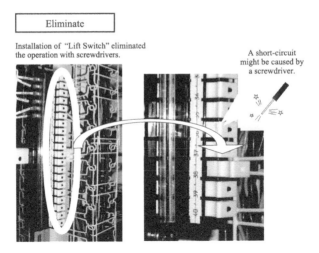

FIGURE 7.11   An instructional material of GUIDE with an example of measure.

on an individual's ability or sense, and that the prevention of errors should precede the mitigation of effects. Note that the same preventive measures can differ in grading depending on their means. For instance, ensuring the proper execution of actions by means of a sensor and interlock system will correspond to eight points ("force the right way of doing things"), whereas it will be given only one point if done by self-check ("notice own error"). Difficulty in execution typically contains cost, leading time, and applicability. This should be considered before deciding on a measure, whereas residual risks and side effects should be considered afterwards. Examination of these topics encourages a risk-oriented and comprehensive evaluation of preventive measures.

This kind of evaluation can be used to decide the priority among preventive measures to be implemented. Some comments on this decision are illustrated in the following:

- Basically, a measure with a higher number of points for effectiveness gets a higher priority for implementation. This decision should ensure that the measure effectively cuts off causality relations among the events. If only one measure with a small number of points for effectiveness is proposed,

TABLE 7.1
Effectiveness Points by Classification Used in GUIDE

Elimination	10
Force the right way of doing things	8
Make things easier to understand; make things easier to do	4
Detect errors; prepare for damages	2
Perceive surroundings; predict unsafe conditions; prioritize safety; maintain ability; notice own error	1

it is recommended to look for another measure with a higher number of points, or to combine the proposal with other preventive measures.

- For cases in which a measure with a smaller number of points for effectiveness is implemented, the residual risk should be made as clear as possible, and its effect should periodically be verified.
- Many preventive measures produce only a small effect. An effective and simple prevention of the recurrence of events can be ensured by making a proper background factors causality diagram and by taking preventive measures with a high score of effectiveness against the causality relations found by the analysis.

STEP 7: IMPLEMENT PREVENTIVE MEASURES

In step 7, we implement the preventive measures that were prioritized in Step 6 by first building a concrete and detailed plan and then carrying it out. It is important to be clear about who is responsible for planning, preparation, and execution. The effects of step 7 are evaluated in step 8. Steps 7 and 8 go beyond a desk analysis, and the particular manner in which they should be carried out has not yet been established.

STEP 8: EVALUATE THE EFFECTS

Finally, the actual effects of the preventive measures taken should be evaluated after their implementation. Before the evaluation, we should verify that the preventive measures were definitely and properly executed. This evaluation will be performed by considering two sets of consequences: those related to the prevention of a recurrence of events and those related to the side effects. The former can be both quantitative and qualitative, such as a decrease in the number of events or a subjective improvement of the easiness of work, for instance. There will always be both good and bad side effects. Some examples are synergy from the improvement of surroundings, increasing busyness, or new types of problems such as automation-induced surprises. Quantitative effects of event prevention may be difficult to evaluate statistically because of the low rate of occurrences, and qualitative evaluations are therefore important.

PRACTICAL EXPERIENCES AND ACTIVITIES FOR PERSONNEL TRAINING

SAFER becomes truly useful only after it has been integrated in the work site. The HFG has therefore organized in-house training in cooperation with the administrative section of some divisions and the training center in TEPCO. The number of such training events has exceeded one hundred in just the last three years. In addition to the standard training course for beginners, we developed an advanced course in August 2007 to train knowledgeable and skilled seniors and instructors. We have received extensive feedback both directly from the training activities and indirectly from analysis results performed on-site. Favorable feedback includes an increase in physical rather than immaterial countermeasures, presentation of the logical basis for preventive measures based on fact-finding, and affirmative comments such as the "notion and interpretation of human errors have changed." Unfavorable feedback has been used as a means for improvements, and some are still being worked on; for instance, that prioritization of various know-hows is unclear or that preventive measures for organizational factors are difficult to grade by the classification used in GUIDE.

TEPCO's nuclear power division and the HFG are planning to train analysts and advisors in all three of our power stations as well as in the headquarters. The former group performs analyses and has considerable knowledge of each department, whereas the latter advise the former from a cross-sectional point of view. During fiscal years 2005 through 2007, the courses were organized fifteen times and approximately one hundred analysts and ten advisors were trained. Besides these courses, we developed three cases of e-learning as instructional material, used to teach the basic notions of human factors engineering and practical guidance that are part of SAFER. At the same time we tried to incorporate the analysis into the daily management of incompatible events. In December 2007, a scheme was introduced so that a root cause analysis (RCA) was requested for each significant event, with a regulatory body evaluating the result. TEPCO has adopted SAFER as their RCA method, although SAFER is not a root cause analysis in a strict sense. The strict RCA idea presumes that true root causes exist, as causes without which the events would not have occurred. This assumption is incompatible with many actual accidents for which various causality relations among background factors have been found to explain the final consequences. We therefore think that the strict RCA idea is not adequate for practical use, and we accordingly use the term "root cause" in more general interpretation. Our emphasis in practice is on "cause" rather than on "root," which means that the whole set of background factors in itself should be understood as the root cause. Based on this idea, we will continue the previously mentioned training activities to promote our systematic ability to perform RCA.

CONCLUDING REMARK: BEYOND PROCEDURES

In autumn 2006, the author had the opportunity to take some courses at the training center of the National Transportation Safety Board, the specialized agency for accident investigation in the United States. During the course, it became clear from many discussions among the participants, most of whom were professional accident investigators, that the most important attitude for accident investigators is to look carefully at the scene and thoroughly at the facts without any preconceptions or hypotheses.

Many efforts have been made to categorize background factors, to prescribe viewpoints, and to standardize the procedure of incident/accident analysis, especially in the nuclear power industry where demands for safety and reliability remain very high. These activities are valuable in order to minimize the difference in results among analysts and to promote a systematic capability on the whole. Yet the problems that come from making assumptions about causes, and the persistence in relying on detailed categories or procedures, cannot be dismissed. In making an analysis, a neutral attitude toward facts, logical thinking, and a consideration of "what induced the results?" are more important than obedience to methodology, one reason being that we are unable to foresee all possible conditions.

ACKNOWLEDGMENTS

All our activities concerning SAFER were conducted by Group Manager Mr. Hideo Sakai, Ex-Manager Mr. Ryutaro Kawano, and many co-operators on-site and at headquarters as well as colleagues in HFG. The author would like to express hearty thanks to all of them for their help and encouragement.

REFERENCES

Aviation Safety Network (n.d.). ASN Aircraft accident Boeing 747-446D JA8904 off Shizuoka Prefecture [WWW page]. URL http://aviation-safety.net/ database/record. php?id=20010131-3

Cabinet Office (2005). White Paper on Traffic Safety in Japan: Traffic Accidents and Safety Measures in FY 2004 [Online Document]. URL http://www8.cao.go.jp/koutu/taisaku/ h17kou_haku/tswp2005web.pdf

Hawkins, F. H. (1987). *Human factors in flight*. Aldershot, UK: Gower Technical Press.

Human error wo Fusegu Gijutsu is the original published title in Japanese. An English translation would be "Techniques to prevent human errors."

Kawano, R. & TEPCO Human Factors Group (2006). *Human error wo Fusegu Gijutsu*. Tokyo: JMA Management Center (in Japanese). The Japan Times Online (2000). Patient death probed as malpractice [WWW page]. URL http://search.japantimes.co.jp/cgi-bin/ nn20000308a6.html

Sansone, C., Morf, C. C. & Panter, A. T. (2004). *The Sage Handbook of Methods in Social Psychology*. Thousand Oaks, CA: Sage Publications

Yoshizawa, Y. (1999). Activities for on-site application performed in Human Factors Group. *Proceedings of the 3rd International Conference on Human Factor Research in Nuclear Power Operation (ICNPO-III)*. Mihama, Japan.

8 A Regulatory Perspective on Analysis Practices and Trends in Causes

Maomi Makino, Takaya Hata,
and Makoto Ogasawara

CONTENTS

INTRODUCTION

To maintain a high level of safety and reliability of nuclear power facilities, it is necessary to reinforce both "soft" or nonstructural aspects of safety design and management and "hard" or structural aspects, such as facilities and equipment.

In recent years, in a number of cases, organizational inadequacies, including a fragile safety culture or the lack of the safety culture itself, have played a role. This can be seen from a series of events and accidents, such as the critical accident at the uranium processing plant of JCO (Nuclear Safety Commission of Japan, 1999), the false notification by Tokyo Electric Power Company (Nuclear and Industrial Safety Agency, 2002), the secondary system pipe rupture at Mihama Nuclear Power Plant Unit 3 of Kansai Electric Power Company (Nuclear and Industrial Safety Agency, 2005), and the criticality accident and unexpected dislodgement of control rods at Shika Nuclear Power Station Unit 1 of Hokuriku Electric Power Company (Nuclear and Industrial Safety Agency, 2007a). Taken together, these events (all described later in this chapter) have called into question the safety and reliability of the entire nuclear power industry.

In response to these circumstances, the Japan Nuclear Energy Safety Organization (JNES) is working very hard to ensure the safety and reliability of the Japanese nuclear power operation and technically to support the Nuclear and Industrial Safety Agency (NISA). To be specific, the JNES is collecting, organizing, and reviewing various information about nuclear incident and accident cases both inside and outside of Japan, looking particularly for factors that are harmful to human and organizational safety and reliability, in order to develop guidelines for judgment that appropriately match the current situation in Japan.

This chapter presents a regulatory perspective on analysis practices and trends in causes, and it describes the intention behind an enacted "Guideline for Regulatory Agencies in Evaluating Contents of Root Cause Analysis by Operators." The objective is to provide a guideline on how to verify the appropriateness of the corrective and preventive actions implemented by operators, based on a root cause analysis of events. This guideline takes four points into special consideration for adequate application. They are:

1. Encouragement of further activities of the operators
2. Flexible interpretation of the intention
3. Versatility of the analysis methods
4. Concepts and consideration of no-blame culture

Moreover, as a viewpoint for regulatory agencies, the guideline places special emphasis on ensuring the neutrality of the analyzing party, the objectivity of analysis results, and the logic of the analysis method.

TRENDS OF CAUSAL HUMAN FACTORS OF REPORTED INCIDENTS IN JAPAN

From the regulator's point of view, the collection, analysis, and evaluation of operational safety related events, as well as feedback to the regulation, are important both to prevent the recurrence of the same events and to prevent the occurrence of similar events. To draw the necessary lessons and take appropriate and timely corrective actions, we must identify the precursors of accidents from operating experience in a systematic way.

Moreover, lessons learned from the operating experience should be exchanged internationally. According to the laws and notifications, the electric utilities in Japan are obliged to report promptly to NISA about incidents and failures occurring in commercial nuclear power stations. Recently, most events occurring in nuclear power plants have had a significant human factors contribution and should therefore be analyzed from this point of view. This matches the trend seen internationally, that the relative portion of human impacts is increasing whereas the technical impact is decreasing. Analysis of the operating experience with human errors therefore becomes important.

DEFINITION OF HUMAN ERROR AND THE CLASSIFICATION SYSTEM OF CAUSAL HUMAN FACTORS

JNES defines human error as a deviation from the requested standard of design, manufacturing, installation, operation, maintenance, or other related standards, as well as from generally requested levels. This is similar to the definition provided by the Institute of Human Factors (IHF) of the Nuclear Power Engineering Corporation (NUPEC).

The classification system of causes of errors to be analyzed by JNES is shown in Figure 8.1. This classification system was originally developed by IHF and corresponds to the human error classification scheme adapted from Rasmussen (1981). Based on an information processing model for human performance, it provided a multifaceted classification scheme consisting of error modes, error mechanisms, and causes of error or situational factors causing error (IHF, 1998; Hasegawa & Kameda, 1998; Monta, 1999; Yukimachi & Hasegawa, 1999). Its development began in 1990 and was promoted on the basis of thorough discussion in the human factors subcommittee established within the IHF. This system was validated by analyzing 199 human error events for 49 nuclear power plants in Japan. These events were selected from 863 incidents and failures reported to the government from fiscal year 1966 through 1995, using the previously mentioned human error definition and criteria for extracting human error incidents and failures. The organization, function, and accumulated knowledge of IHF were placed under the control of JNES in 2003.

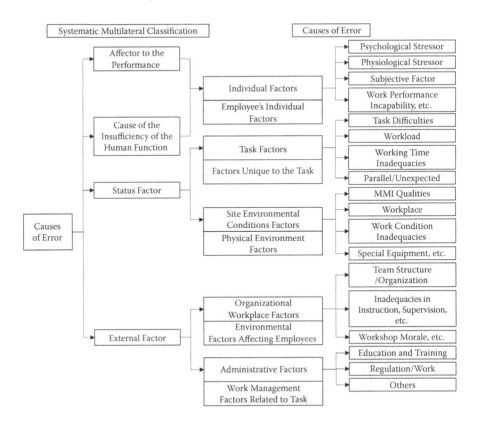

FIGURE 8.1 Classification system of human factors for JNES analysis.

Since then, JNES has continuously analyzed incidents and failures using the same classification system.

In the classification scheme of human error, the *causes of error* or *causal factors* are critical elements for the purpose of exploring causes of human error occurrence synthetically and structurally. Based on experience and knowledge up to now, and on the results of analysis and evaluation of events, etc., causes of error occurrence can be classified by five major characteristic factors:

1. *Individual characteristic factors*, such as psychological stressors, physiological stressors, and work performance and capability
2. *Task characteristic factors*, such as task difficulties, workload inadequacies, and parallel/unexpected tasks
3. *Site environmental conditions characteristic factors*, such as man–machine interface (MMI) inadequacies, workplace inadequacies, and work condition inadequacies
4. *Organization characteristic factors*, such as team structure/organization, inadequacies in instruction, and supervision and workshop morale
5. *Administrative characteristic factors*, such as education and training, regulation, and work planning

The chart in Figure 8.1 is a help in reviewing the causes triggered by important performance-shaping factors (PSFs) that change the likelihood of error. One can select many factors that are closely related to each of the characteristic factors.

TRENDS IN CAUSES

The human factors database was established in 1998 and has been maintained since, first by IHF and since 2003 by JNES. The database has registered 498 human error events in all. To ensure continuity and the same quality of stored data, JNES has used the same method, criteria, and requirements for human error analysis and database as IHF. During a validation study of the previously mentioned classification system of causes of errors, human error events up to 1995 were analyzed in bulk based on incident reports. Since then, human errors from the previous fiscal year have every year been analyzed and registered in the database. Altogether, 257 human error events in Japan are registered among 1,096 incidents and failures reported to the government from fiscal years 1966 through 2007. In addition, 53 other human error events in Japan and 188 human error events in overseas countries are registered as references of important operating experiences. The reporting rules were, however, changed in 1981. In order to keep the same criteria in the comparison of trends in causes, 193 human error events in commercial operating nuclear power plants, which were reported as required by the laws, and the notification within a period from fiscal years 1981 to 2005, and registered in a JNES database, were therefore subjected to the human factors analysis of trends in the causes (Makino, 2007).

Figure 8.2 shows the dominant cause among five major characteristic factors which led to 193 human error events. Individual characteristic factors are most dominant, accounting for 249 factors or 36% of all human error events, followed by 175

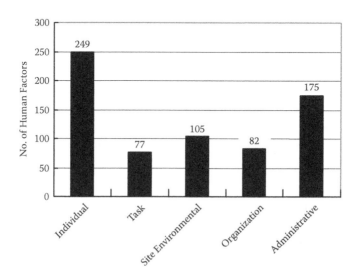

FIGURE 8.2 Number of major characteristic human factors extracted by the analysis of human error events.

TABLE 8.1

Number of Human Factors Extracted by the Analysis of Human Error Events

Five Major Characteristic Factors	Dependent Elements	Extracted Number
Individual characteristic factors	Psychological stressors	27
	Physiological stressors	10
	Subjective factors	117
	Work performance incapability	94
	Others	1
Task characteristic factors	Task difficulties	31
	Workload inadequacies	20
	Working time inadequacies	11
	Parallel/unexpected tasks	14
	Others	1
Site environmental conditions characteristic factors	Man–machine interface (MMI) inadequacies	58
	Workplace inadequacies	40
	Work condition inadequacies	7
	Special equipment	0
	Others	0
Organization characteristic factors	Team structure/organization	27
	Inadequacies in instruction and supervision	52
	Workshop morale	3
	Others	0
Administrative characteristic factors	Education and training	15
	Regulation/work planning	159
	Others	1

administrative characteristic factors covering 25%, and 105 site environmental characteristic factors covering 15%.

The five major characteristic factors are subdivided into further detailed elements for each factor, as shown in Table 8.1. The most dominant element is "regulation/work planning, accounting for 159 or 23% of total number of extracted elements," which belongs to the administrative characteristic factors group. Other important elements are "subjective factors, accounting for 117 and covering 17%" and "work performance incapability, accounting for 94 and covering 14%," which belong to the individual characteristic factors group; "MMI inadequacies, accounting for 58 and covering 8%," which belongs to the site environmental conditions characteristic factors group; then "inadequacies in instruction and supervision, accounting for 52 and covering 7%," which belongs to the organization characteristic factors group.

Table 8.2 shows a comparison of human error occurrences under the specific work types in nuclear power plants. It can be seen that the human errors cause under maintenance work accounts for 43%, that the human errors cause under construction

TABLE 8.2
Comparison of Human Error Occurrence Under Specific Work Types

Specific Work Type	Details	Human Error Occurence	Specific Work Type	Details	Human Error Occurence
Operation	Startup	5	Construction	Design	11
	Power operation	2		Fabrication	20
	Periodical inspection	6		Construction	10
	Abnormality check	4		Others	1
	Shutdown	2	Technical	Technical control	0
	Operation during shutdown	9		Radiation control	0
	Commissioning/ special tests	3		Chemical control	0
	Others	1		Burning control	0
Maintenance	Work preparation	1		Nuclear fuel inspection/ refueling	6
	Overhaul	16		Others	0
	Repair and parts replacement	8	Administration	Rules/ standards	5
	Assembling	30		Manuals/ drawings	20
	Abnormality check	3		Plans/work schedule	1
	Calibration	7		Record control	0
	Test/inspection	8		Inst/comm. between organizations	2
	Cleaning/clear away	4		Education/ training	0
	Others	6		Others	1
			Others	Others	1

work accounts for 22%, that the human errors cause under operation work accounts for 17%, and that the human errors cause under administration work accounts for 15%. In the maintenance work, the human errors cause by assembling work is most conspicuous, followed by the human errors caused by manuals/drawings in administration work and by fabrication in construction work.

To determine a trend or changes over time, one can refer to Figure 8.3, which shows the chronological trend in the comparison of five major characteristic factors. The ratio

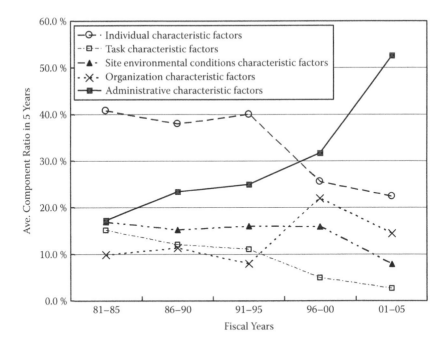

FIGURE 8.3   The chronological trend in the comparison of five major chracteristic factors.

of causes due to administrative characteristic factors increased remarkably after the period of 1991 to 1995. This category occupies the majority (53%) among five major characteristic factors in 2001 to 2005. On the other hand, the ratio of causes due to individual characteristic factors tended to decrease in and after the period of 1996 to 2000. Motivated by this trend, the exhaustive investigation into the root causes of nonconformity due to organizational factors underlying human error events will be enforced in order to decrease human error events in nuclear power plants.

The bar graph in Figure 8.4 indicates the number of events per operating unit, subject to reports, and a line graph indicates the changes in the ratio of human error events to the reported events. After 1981, the number of reported events per operating unit shows a generally decreasing trend. After 2001, the number mostly stayed around 0.5 or below per operating unit despite some decrease and increase.

The ratio of human error events to the reported events ranged between 15 and 40% from 1981 to 2004, with an average of 22.8%. But in 2005 the ratio jumped up to 73%, almost double the highest ratio of the previous period (39% in 2003). It is important to determine whether this was just a temporary phenomenon or whether it will continue in the future. This will be done by continued data collection and analysis and by judging the trend on a yearly basis. However, the conditions in our society, as well as the conditions for a nuclear industry in Japan, have changed. This has led to an increase in the perceived importance of human and organizational factors and a decrease in human error events in nuclear power plants, as shown by discussions and recommendations by committees and subcommittees of a regulatory agency (Nuclear and Industrial

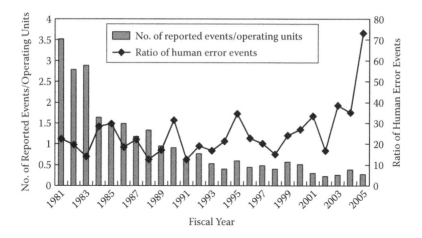

FIGURE 8.4   Number of reported events per operating unit and ratio of human error events to number of reported events.

Safety Agency, 2006). Additionally, the relative contribution of technical causes to the reported events has decreased due to continuous implementation of improvements in nuclear power plants through the years. This, together with the increase in perceived importance, could explain why more details of human and organizational factors are reported and the transparency of causes for human errors has increased.

SUBSTANTIAL IMPACT OF RECENT ORGANIZATIONAL INCIDENTS ON SOCIETY

As shown in Figure 8.4, the nuclear industry in Japan maintained excellent safety performance during the 1990s on a national and international level. Although the frequency of occurrences has been extremely low, recent incidents have had a substantial impact on our society. The characteristics of recent incidents show an influence from organizational systems in addition to the inevitable failures and troubles of equipment or human errors.

The JCO Accident

In 1999, a criticality accident occurred at JCO's uranium processing plant (Nuclear Safety Commission of Japan, 1999). During uranium refining work, the uranium nitrate solution in a precipitation tank achieved a critical state, which resulted in radiation exposure for fifty-nine JCO personnel and at least seven residents in the vicinity, as well as the death of two JCO employees. The report by the investigation committee for the criticality accident, issued in December of 1999, indicated that risk perception of a criticality event had faded with time and was eventually lost, and that deviations from accepted standards had escalated in excessive pursuit of working efficiency. The report recommended:

- A thorough system to ensure safety by process control and work management shall be established, and appropriate risk projection and risk management shall be made routinely and appropriately as the operator's responsibilities

- Establishment of each an engineer's consciousness and ethic
- Further consideration of human factors
- Efforts to establish a nuclear safety culture

The TEPCO Case

In 2002, a case of falsified self-controlled inspection records by the Tokyo Electric Power Company (TEPCO) was revealed. NISA investigated a suspicion that the operator had made false records about the detection, repair, etc., of cracks in its three nuclear power stations from the late 1980s to the 1990s. The cracks were discovered in self-imposed inspections conducted by a contractor at the request of the operator. In the process of those investigations, the operator submitted a list of twenty-nine cases with suspicion of dishonest acts in the three power stations involved. This revealed that the components about which false entries were suspected to have been made in self-imposed inspection records were shrouds, shroud head bolts, steam dryers, jet pumps, and in-core monitor housings, among other things, and that false entries were suspected to have been made in the inspection results, repair records, dating, and so forth, concerning cracks or indications of cracks (Nuclear and Industrial Safety Agency, 2002). The interim report on the falsification of self-controlled inspection work records, etc., issued by NISA in October of 2002, indicated:

- The nuclear power division of the licensee had built a kind of unique "territory" mainly by engineers engaged in nuclear business, which had fostered an atmosphere that made it hard for outsiders of the division to get involved in the business. Therefore, decision making for failures, repairs, and the safety of facilities had, in practice, been made by a limited number of persons who consisted mainly of technical specialists. It was judged that audits of the decision processes and the evaluation of the results, by sections other than the nuclear power division, including top management, had not been carried out adequately.
- Thus, for this case it was considered as a fundamental cause that recognition of operational importance of quality assurance had not been established throughout the organization, and that the licensee's companywide quality assurance functions for activities of the nuclear power division had "gone to sleep."

The KEPCO Accident

In 2004, a pipe rupture accident happened in the secondary system at Unit 3 of the Mihama Power Station of the Kansai Electric Power Company (KEPCO). The pipe rupture was caused by reduced pipe strength, which was due to wall thinning by so-called erosion/corrosion at the downstream side of an orifice in the secondary system piping. The direct cause of the accident was that the effects of erosion and corrosion were overlooked for many years, because the pipe in question was mistakenly missing from the inspection list. Further investigation revealed that the root causes of the accident were the poor maintenance management and insufficient quality assurance systems of the operator, with the background of the operator's declining safety culture (Nuclear and Industrial Safety Agency, 2005). At that time, five persons were killed and six persons were injured due to steam and/or high temperature water flowing from the ruptured pipe. The report indicated that:

- Both the operator and a contractor neglected their duties to check the inspection lists and to establish good communication. They failed to correct the mistakes in the inspection list that had been initially created by the contractor.
- The operator improperly managed the subcontractors. The item in question had been mistakenly left off the inspection list.
- The operator held a biased attitude toward process operations. At some point, noncompliance with technical standards was accepted as normal.

The HEPCO Accident

In 2007, a criticality accident and unexpected dislodgement of control rods at Unit 1 of the Shika Nuclear Power Station, operated by the Hokuriko Electric Power Co. (HEPCO), were discovered. The incident had occurred June 18, 1999, during a period of maintenance outage. The cover-up of the accident and data falsifications were not disclosed until eight years later.

In an attempt to arrange the line-up for a specific validation test, four workers in the plant, in turn, isolated the hydraulic control unit of the control rod hydraulic system in parallel, without close communication with one another and without following the work procedure. This caused the control rods unexpectedly to dislodge in a withdrawal direction. The reactor went into a critical condition and scram signals were initiated by the neutron monitoring system. However, the control rods were not inserted at once, because the necessary valves for rod insertion were closed, and because the related hydraulic control unit accumulators were not pressurized due to the preparation of the specific validation test.

The NISA investigation report indicated the following as its assessment of the organization on ensuring safety (Nuclear and Industrial Safety Agency, 2007a):

- Although the management of the station recognized that critical accidents must be reported to the government, as provided by law, they did not report it because keeping the construction schedule of Shika Nuclear Power Station Unit 2 was given first priority. The chief engineer of reactors forgave the wrong decision-making by the site superintendent and did not discharge his rightful duty from the standpoint of an independent technical position.
- In order to conceal the critical accident, various records including legal records were falsified, and the correct recordings were not given out.
- The manager of the operating section instructed that the related description of the accident not be recorded on the shift log.

APPROACH TO ESTABLISHING A GUIDELINE AND A STANDARD FOR A ROOT CAUSE ANALYSIS

Up to the present, root cause analysis has been enforced by licensees as part of the self-controlled operational safety activities comprising corrective and preventive actions provided by the rules of quality assurance. However, the licensees' efforts have not been sufficient to rectify the shortcomings of the conventional method. Often, the licensees' approach to correcting nonconformance has been superficial; that is, it has been directed at the improvement of manifest events only, whereas the activities to analyze and improve the root cause, centered on organizational causes such as an inappropriateness of the management system, have not been adequately performed so far. Because of this, there have been frequent accidents and problems partly associated with organizational causes, the root causes of which have remained unaddressed.

Although the nuclear industry seems to have attained its maturity, the developments mentioned here make it clear that the industry should take thorough corrective and preventive actions. The basis for this should be the root cause analysis, through which the latent organizational factors for each event are made clear, in order to make sure that they do not recur.

As a consequence, the process and the system for implementation of a root cause analysis have been defined in the regulatory rules, as provisions of quality assurance in the operational safety program, and the licensees have been forced to include them in their quality assurance programs. In addition to this, NISA has provided regulatory requirements for the evaluation of how a root cause analysis is implemented by a licensee, as well as the requirements for the process of a root cause analysis in the rules. All licensees have been notified of these developments.

The regulatory requirements comprise the following four items:

1. The implementation of root cause analysis shall ensure the neutrality of the analysis, the objectivity of the analysis result, and the logic or consistency of the analysis method.
2. For events that have a significant impact on safety, appropriate corrective and preventive actions shall be carried out, and a root cause analysis for each event shall be implemented to ensure prevention of a recurrence.
3. Concerning other events that do not have a significant impact on safety, analysis of the accumulated data related to nonconformance shall be conducted after taking corrective actions, and a root cause analysis shall be implemented depending on the necessity to implement preventive actions.
4. Corrective and preventive actions should be based firmly on the result of the root cause analysis, and a specific implementation plan should be clarified and conducted without fail.

The implementation procedure of root cause analysis is shown in Figure 8.5.

Based on these requirements, the regulatory guideline for inspectors to evaluate the results of a root cause analysis implemented by a licensee was established in December of 2007. The regulatory guideline was developed via discussion by a subcommittee on a draft of a guideline worked out by NISA and JNES (Nuclear and

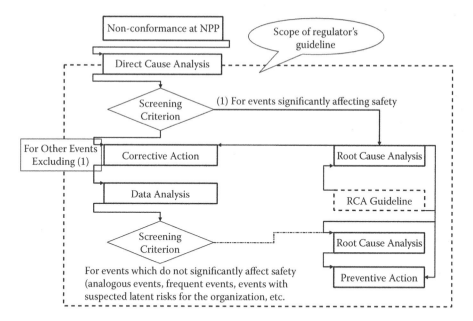

FIGURE 8.5 Process flow for enforcement of root cause analysis.

Industrial Safety Agency, 2007b). The guideline is named *Guideline for Regulatory Agencies in Evaluating Contents of Root Cause Analysis by Operators*. The Japan Electric Association (JEA) established the association standards (JEA, 2007) relating to a guide for implementation of a root cause analysis. These standards provide an adequate system, methods, screening, reporting, measures, effect evaluation of measures, etc., relating to implementation of a root cause analysis. NISA has evaluated this and approved it as the standard to meet the regulatory requirements.

The enforcement of this regulatory guideline is expected to encourage the prevention of incidents stemming from organizational systems, due to thorough implementation of corrective and preventive actions. The following effects are expected from the enactment of this guideline:

- The definition of terms such as "root cause analysis," "organizational factors," etc., will promote a common understanding between regulators and licensees.
- The provision of minimum requirements for events subject to implementation of a root cause analysis will encourage positive investment of resources in a root cause analysis and stimulate adequate common information for preventive actions among licensees.
- The provision of a requirement to ensure the neutrality of a setup for a root cause analysis is expected to ensure the reliability of results.
- The provision of a requirement to ensure objectivity in implementing a process of root cause analysis is expected to help identify or extract adequately organizational factors that are latent in a target event and to draw up adequate and substantial measures for the extracted organizational factors.

- The provision of a requirement to ensure the logic of the analysis method will lead to a more systematic account of organizational factors.
- The provision of a guide for evaluating the appropriateness of corrective and preventive actions is expected to help ensure secure checks on the status of implementation and effectiveness of actions.

CONTENTS OF THE REGULATORY GUIDELINE

The regulatory guideline (hereafter referred to as just "guideline") provides the guidance to verify that the corrective and preventive actions implemented by a licensee, based on the results of a root cause analysis, are appropriate. The guideline takes four points into special consideration for adequate application:

1. In addition to the judgment of whether or not the root cause analysis satisfies the government requirements, an evaluation should be made with the aim to encourage further activities of the operators to improve the methodology, process, and results of the root cause analysis.
2. In the event of any doubt about descriptions in the guideline, a flexible interpretation of the intention of the root cause analysis should be implemented rather than adhering to the specific wording.
3. When verifying the licensees' approach, positive discussion should be held with licensees on a continuous basis. Versatility of the analysis methods and concepts adopted by licensees should be allowed.
4. The party concerned should have sufficient awareness of the fact that there are various factors in the behaviors of the personnel involved in a nonconformance. In addition to the negative factors, such as misunderstanding, wrong judgment, and insufficient confirmation, there may be negative effects (influences) caused by excessive implementation of actions based on the expectation for positive effects, such as improvement of the working environment, efficiency improvement, and the pursuit of cost reduction.

Definition of the Terms

Terms such as "root cause analysis" and "organizational factors" are sometimes used with a wide variety of meanings by operators and others. The terms used in the guideline are therefore defined, to ensure a clear mutual understanding between the regulatory body and operators:

Root cause analysis: To analyze organizational factors based on direct cause analysis and to take actions to improve the management system.
Note: Although the analysis of technical factors is included in this category in general, the term is defined as above to give consideration to the fact that many accidents and failures will occur due to lack of appropriate organizational response rather than to already known technical factors.
Direct cause analysis: To take corrective and preventive actions by implementing direct factor analysis on the concerned accident/failure or nonconformance event.

Direct factors: Factors in a local process associated with the occurrence of an event. They are the factors that constitute the direct cause of the equipment damage or human error. Direct factors include both technical factors and human factors.

Organizational factors: A group of factors related to organizational activity that failed to prevent a direct factor ahead of time.

Human errors: Human behaviors related to design, fabrication, installation, operation, maintenance, management, etc., that deviate from the requested standard.

Human factors: A group of all the factors surrounding human beings related not only to the factors associated with human characteristics, but also to workplace environments, job environments, job characteristics, and management characteristics.

Guides to Verify Process and Results of Root Cause Analyses Implemented by Operators

Based on the regulatory requirements for root cause analysis, three additional guidelines were developed to confirm the process and the result of root cause analyses:

1. The guideline to confirm that the analyzing party is neutral
2. The guideline to confirm that the analysis results are objective
3. The guideline to confirm that the methodology used for analysis is logical

It has further been made clear that:

When applying the contents of the following descriptions, the guide for application and its depth should be judged based on the importance of each item seen from the analysis results and operator's management system, instead of applying all the items in a uniform manner.

In other words, the guideline should be applied carefully and not in a routine or rule-based manner. An overview of each guideline is given in the following.

The Guideline to Confirm that the Analyzing Party is Neutral

For an accurate analysis implementation, the neutrality of the analyzing party and nonsuffering disadvantage in personnel evaluation must be assured. Also, to extract organizational factors, the interview with the manager is indispensable.

This leads to the following four guidelines:

1. The analyzing party shall comprise the people representing the functions not directly involved in the event concerned.
2. Access to the essential data shall be authorized. Further, implementation of research, including interviews with management and the related functions, shall be ensured.
3. The individual who implemented the root cause analysis shall be protected from potential disadvantageous treatment associated with the analysis and its results.

4. The individual who is in charge of the root cause analysis shall have experience in safety preservation activities in power plants, or shall understand such practice in addition to experience or education/training related to the root cause analysis.

The Guideline to Confirm that the Analysis Results are Objective

To elaborate this guideline, the following five precise guidelines are introduced:

1. In the contents of events and problems, the concerned functions and individuals shall be kept anonymous and the behaviors concerned shall be described in detail.
Note: "Identification based on the anonymous basis" refers to identification based on one's responsibility, authority, and role in an organization. If multiple individuals have an identical responsibility, authority, and role, they shall be identified with symbols such as A and B.
2. Problems shall be clarified and described quantitatively as much as possible.
3. Organizational factors corresponding to the problem shall be clarified and described in detail.
4. Actions corresponding to organizational factors shall be clarified and described in detail.
5. For improved understandability, the specific example of each guideline is specified in the guideline.

The Guideline to Confirm that the Methodology Used for Analysis is Logical

For this guideline, the following six precise guidelines are introduced:

1. The root cause analysis shall systematically consider the viewpoints of organizational factors and their causal relationships depending on the reported events. As a reference list of organizational factors, the JNES Organizational Factors List (JOFL) is provided.
Note: Systematic analysis refers to the identification of the factors based on a specific framework and the narrowing of the targeted factors depending on the magnitude of impact on the results. This is done to prevent omission of any important factor to prevent recurrence of accidents caused by similar factors.
2. Trans-sectional analysis of events, data, and research results from various viewpoints shall be conducted as necessary to explore common factors.
3. The analysis shall have sufficient depth to be able to improve inappropriateness of the management system.
4. Depending on the need, the possible inappropriateness of the past corrective and preventive actions shall be reviewed.
5. Depending on the need, difference factors caused by change and modification before and after the event concerned shall be analyzed.
6. Depending on the need, an analysis shall be conducted of whether or not a barrier was present to prevent event occurrence or human error, and whether or not such a barrier was lost or dysfunctional.

The Guideline to Verify Appropriateness of Corrective and Preventive Actions

It is possible that some of the present reports for incidents do not include an actual plan or process of evaluation activities for corrective and preventive actions. To improve this situation, the following six guidelines are introduced:

1. Corrective and preventive actions corresponding to the root cause analysis shall be formulated.
2. If no action is taken, the reason for this shall be indicated clearly.
3. An evaluation of the effects of corrective and preventive actions shall be conducted and the extent of their ability to prevent events caused by a similar direct factor shall be indicated clearly.
4. An evaluation shall be conducted on the side effects associated with the corrective and preventive actions.
5. A specific implementation program of corrective and preventive actions (system, schedule, resources, follow-up method, method of evaluating efficacy, priority, etc.) shall be identified clearly, accepted by the staffs concerned, and feasible.
6. The necessity and applicable range for cross-cutting development of corrective and preventive actions shall be reviewed.

REVIEW OF THE GUIDELINE BASED ON FEEDBACK FROM OPERATIONAL EXPERIENCE

Only two cases have been practically implemented by licensees under the new root cause analysis system that came into force in December 2007. They have been evaluated by NISA in accordance with this new established guideline. Obviously, additional actual use of this guideline will provide both operators and the regulatory body with valuable lessons about root cause analysis. This will also provide essential feedback to the guideline. It is the intention to review the guideline continuously in the future.

DEVELOPMENT OF JOFL

The method used by a licensee for a root cause analysis should not be restricted by the regulatory body. A licensee should use a method that is recommended by an association standard (JEA, 2007). But the regulatory body expects licensees to adopt an adequate method and implement it correctly. It is also important that the essential organizational factors from various root cause analyses are considered together so that they can possibly be combined. In order to facilitate this, the JNES has prepared the JNES Organizational Factors List (JOFL) as a reference list for the regulatory body to confirm the appropriateness of organizational factors found by the licensees' root cause analyses.

The JNES has developed a safety culture evaluation support tool, called SCEST, to characterize the fragility of safety culture (Makino, Sakaue, & Inoue, 2005; Safety Standard Division, 2006). The JNES has also developed an organizational reliability model (OR model) to identify organizational factors that may be disincentive of safety culture (Institute of Human Factors, 2003). The JOFL integrated these

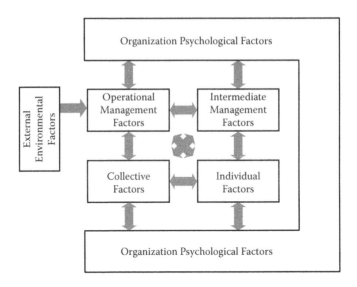

FIGURE 8.6 Potential causal relationships among organizational factors considered in conducting an RCA.

evaluation items with the readjusted organizational factors evaluation items to create a new original list (Safety Standard Division, 2007).

Reflecting the results of the application to the specific cases for which NISA implemented special audits, an original list was processed to produce the JOFL. This reference list is composed of six key factors that refer to a structure of 63 intermediate classifications and 137 viewpoints, as well as questions for the confirmation of each viewpoint. The six key factors are external environmental factors, organizational psychological factors, operational management factors, intermediate management factors, collective factors, and individual factors. They are referred to in the guideline as viewpoints for organizational factors. Figure 8.6 indicates potential causal relationships among organizational factors considered in conducting a root cause analysis.

VIEWPOINTS FOR ORGANIZATIONAL FACTORS IN ROOT CAUSE ANALYSIS

EXTERNAL ENVIRONMENTAL FACTORS

The factors related to the external environment of the organization concerned can be included among the set of organizational factors if the impact of economic status, regulatory response policy, external communication, general reputation, and so forth are important for the issue concerned.

ORGANIZATIONAL PSYCHOLOGICAL FACTORS

These are the factors related to the common sense of value among organization members as a mode of thinking or behavior, formed during a long period in the organization (each collective level such as corporate level, power plant level, function level, group level, and team level). They can be expressed in a form of consciousness, awareness, and behavior. They can be included in the set of organizational factors if they are important for the issue concerned.

OPERATIONAL MANAGEMENT FACTORS

The factors related to the operational management of the head office can be included in the set of organizational factors if the following factors are important for the issue concerned. These factors are illustrative of inappropriateness or lack of specificity or effectiveness of top management commitment, organizational operation (operation status, organization structure, organization objectives and strategies, decision making of head office, etc.), personnel operation, company policies and compliance criteria and standards, communication between the head office and power station, and self-evaluation (or third-party evaluation).

INTERMEDIATE MANAGEMENT FACTORS

The factors related to the management operation of the power plant can be included in the set of organizational factors if the following factors are important for the issue concerned. These factors are illustrative of inappropriateness or lack of specificity or effectiveness of function-level organizational operation (objectives and strategies, establishment of a Quality Management System, improvement of manuals, etc.), conformance to rules, continuous education of the organization (handing down of skills, reflection of operation experience), personnel management, communication, procurement management (communication and control with cooperative companies), human resources management related to organizational structure (role and responsibility, selection and arrangement, performance, education, and training), engineering control, work control, change control (control at modification of the organization, control at change of work, etc.), nonconformance control, corrective action, and documentation control.

COLLECTIVE FACTORS

These are the factors related to the groups at each level of the organization (e.g., management, division, section, team on shift, job team, etc.). They can be included in the set of organizational factors if their negative impacts of inter- or intra-cohort communication, knowledge/education, collective narrow-sight and decision making based on principle of individuality, etc., are important for the issue concerned.

INDIVIDUAL FACTORS

The factors related to the individuals (employees or managers) in the organization or groups can be included in the set of organizational factors if their impacts, such as lack of knowledge or skill, leadership, eagerness/prudence for safety, eagerness for management, consideration of site staffs, motivations, stress, etc., are important for the issue concerned.

The following examples of viewpoints can be used to decide whether or not causal relationships of organizational factors associated with intermediate management factors of the power plant shall be analyzed:

- Whether or not intermediate management factors that caused inappropriate behavior have been analyzed.
- Whether or not operational management factors that caused the intermediate management factors have been analyzed.
- Whether or not the association of inappropriate behaviors, intermediate management factors, and operational management factors has been analyzed in a logical manner.
- Whether or not the association of individual psychological factors (individual factors), workplace psychological factors (collective factors), and organizational psychological factors have been analyzed, depending on the necessity.

CONCLUSION

The chronological trend of reported incidents indicates that the ratio of causes due to administrative characteristic factors increased substantially after the period of 1991 to 1995. This category occupies the majority among five major characteristic factors. On the other hand, the ratio of causes due to individual characteristic factors tended to decrease during and after the period of 1995 to 2000. Motivated by this trend, the exhaustive investigation into the root causes due to organizational factors was enforced in order to decrease human error events in nuclear power plants.

The conditions of our society as well as the nuclear industry in Japan have been changing. The perceived importance of human and organizational factors has increased and the importance of finding ways to decrease human error events in nuclear power plants has been reinforced. At the same time, the relative contribution of technical causes to the reported events has also decreased due to continuous implementation of improvements in nuclear power plants throughout the years. This, together with the greater perceived importance of human and organizational factors, could explain the fact that more details of human and organizational factors are reported and that transparency of causes for human errors has increased recently.

Although the frequency of occurrence of safety-related events has been extremely low, recent incidents have had a substantial impact on our society. It is characteristic of recent incidents that they stem from organizational systems, in addition to the inevitable failures and troubles due to malfunctioning equipment and human errors. These developments have forced the nuclear industry to take thorough corrective and preventive actions. Part of that has been to perform a root cause analysis and to look for the latent organizational factors for each event, in order to prevent recurrence effectively.

The regulatory requirements for the implementation of a root cause analysis ensure neutrality of the analysis, objectivity of the analysis result, and logic of the analysis method. Based on the requirements, a guideline for a root cause analysis has been worked out and enacted in Japan. The guideline indicates the steps to verify that the corrective and preventive actions implemented by a licensee based on a root cause analysis and its analysis results are appropriate.

The guideline makes clear that a root cause analysis shall present a systematic analysis that considers the viewpoints of organizational factors and their causal relationships depending on the reported incidents. The JOFL has been developed as a reference list for this purpose.

ACKNOWLEDGMENTS

The authors would like to express special thanks to the staff of the Nuclear Power Inspection Division of NISA and the members of the study committee of JNES for supplying valuable information and suggestions, and encouraging us to make the guideline, and developing the JOFL.

REFERENCES

Hasegawa, T. & Kameda, A. (1998) *Analysis and evaluation of human error events in nuclear power plants.* Tokyo: NUPEC. (http://www.nupec.or.jp/database/paper/paper_10/p10_human/R10-07-04.htm)

Institute of Human Factors (1998). *Analysis and evaluation of human error events. Annual report on human factors research project 1997.* Report no. IHF/9801. Tokyo: Nuclear Power Engineering Corporation.

Institute of Human Factors (2003). *Making of guideline of understanding and enforcement of safety culture. Annual report on human factors research project, 2002.* Report no. IHF/0301. Tokyo: Nuclear Power Engineering Corporation.

JEA (2007). *Application guide to quality assurance code for safety in nuclear power plants (JEAC4111-2003) — Operation phase of nuclear power plants.* JEAG4121-2005 [2007 Suppl. 2]. Tokyo: Japan Electric Association.

Makino, M. (2007). *The result of human factor analysis of human error events and development of a guideline for a root cause analysis.* Seminar material no. NE-07-8.Tokyo: The Institute of Electrical Engineers of Japan.

Makino, M., Sakaue, T. & Inoue, S. (2005). Toward a safety culture evaluation tool. In N. Itoigawa, B. Wilpert & B. Fahlbruch (Eds.), *Emerging demands for the safety of nuclear power operations* (pp. 73–84). Boca Raton, FL: CRC Press.

Monta, K. (1999). *Significance of safety culture for nuclear power operation.* Tokyo: NUPEC. (http://www.nupec.or.jp/database/paper/paper_11/p11_human/R11-08-04.htm)

Nuclear and Industrial Safety Agency (2002). *The issues on falsification of self-controlled-inspection work records, etc. at a nuclear power station.* Investigation Subcommittee Interim Report. Tokyo: NISA.

Nuclear and Industrial Safety Agency (2005). *Summary of the final report on the secondary system pipe rupture at Unit 3, Mihama NPP, KEPCO.* White papers/reports. Tokyo: NISA.

Nuclear and Industrial Safety Agency (2006). *The improvement of the inspection system for nuclear power generation facilities.* NISA Report. Tokyo: NISA.

Nuclear and Industrial Safety Agency (2007a). *A criticality accident and unexpected dislodgement of control rods at Unit 1 of Shika nuclear power station, Hokuriku Electric Power Co. in 1999.* NISA Investigation Report. Tokyo: NISA.

Nuclear and Industrial Safety Agency (2007b). *Guideline for regulatory agencies in evaluating contents of root cause analysis by operators* (NISA document no. NISA-166c-07-10). Tokyo: NISA.

Nuclear Safety Commission of Japan (1999). *The criticality accident at the uranium processing plant.* Investigation Committee Report. Tokyo: NSC.

Rasmussen, J. (1981). *Human errors: A taxonomy for describing human malfunctioning in industrial installations* (Risø-M-2304). Roskilde, Denmark: Risø National Laboratory.

Safety Standard Division. (2006). *The safety culture assessment method (The implementation manual). JNES-SS report no. JNES-SS-0616.* Tokyo: Incorporated Administrative Agency, Japan Nuclear Energy Safety Organization.

Safety Standard Division. (2007). *Preparation of regulatory requirements for cultivation and infiltration of nuclear safety culture in organizations.* Annual report no. 07-KIHI-HOU-0003. Tokyo: Incorporated Administrative Agency, Japan Nuclear Energy Safety Organization.

Yukimachi, T. & Hasegawa, T. (1999). *Analysis of human errors occurred at nuclear power plants in Japan.* Tokyo: NUPEC. (http://www.nupec.or.jp/database/paper/paper_11/p11_human/R11-08-10.htm)

9 An Ironic Outcome of Thorough Accident Prevention

Masayoshi Shigemori

CONTENTS

RAILWAY ACCIDENTS IN JAPAN

Different industries implement a range of measures to reduce accidents. Examples in the aviation industry are the traffic alert and collision avoidance system (TCAS), as a technology-based measure, and cockpit resource management (CRM) training, as a human factors measure. In the railway industry, examples are automatic train stop (ATS) as a technology-based measure, and pointing-and-call checking, as a human factors measure. In nuclear power plants the emergency core cooling system (ECCS) illustrates another technology-based measure. The purpose of this chapter is to describe problems relating to accident prevention in Japan. However, the type and degree of events considered to be accidents differ from industry to industry. We, therefore, need to define accidents before discussing them.

In general, we define accidents not only as abnormal events that somehow have origins in the human–technology–organization complex, but also by the severity of the loss (Figure 9.1) and the threshold of acceptable loss as decided by organizational policy or social values. Some organizations consider a train delay of a few minutes an accident, whereas others do not consider even a few hours of delay as an accident (Figure 9.2).

The railway industry in Japan sets stricter thresholds for accidents than other industries, such as the airline business (Figure 9.3). Although it is difficult to compare organizational policies for safety among industries, Japanese railway companies, for example, administer train schedules to the nearest second. The mean delay for Shinkansen trains was 0.3 minutes in 2006 (Central Japan Railway Company, 2007); in contrast, although the mean ratio of on-time flights for major airline companies in

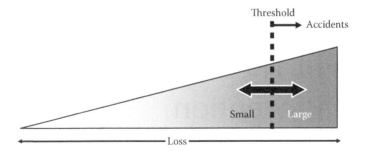

FIGURE 9.1   Role of severity of loss on definition of accident.

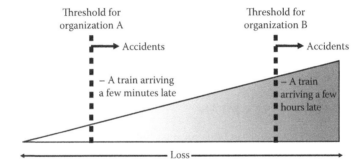

FIGURE 9.2   Different thresholds of accidents in different organizations.

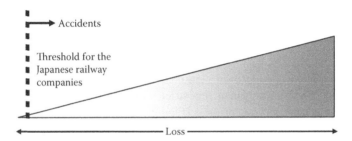

FIGURE 9.3   Strict definition of accident by Japanese railway companies.

TABLE 9.1
Types and Definitions of Railway Accidents in Japan

Type of Accident	Definitions
Train collision	An accident in which a train collides with or touches another train or carriage
Derailment	An accident in which a train is derailed
Train fire	An accident in which a train catches fire
Level crossing accident	An accident in which a train or carriage hits or touches a pedestrian, car, etc., on a level crossing
Railway accident involving traffic hindrance	An accident in which a train or carriage hits or touches a pedestrian, car, etc., on a road other than a level crossing
Railway accident involving casualties	An accident involving casualties (i.e., fatality or injury) caused by train or carriage operation, excluding the above accidents
Railway accident involving material damage	An accident involving material damage amounting to more than five million yen, excluding the above accidents

Japan was 93.20% in 2006, this figure includes flights that departed with delays of less than fifteen minutes (Ministry of Land, Infrastructure and Transport, 2008).

Accidents on railways in Japan are divided into railway accidents and transport disorders by the Regulations for Railway Accident Reporting (Ministry of Land, Infrastructure and Transport, 1987). Railway accidents are further divided into train collisions, train derailments, train fires, level crossing accidents, railway accidents with traffic hindrance, railway accidents with casualties, and railway accidents with material damage (Table 9.1). In addition to these, railway companies must inform the District Transport Bureau of events involving crew or passenger deaths; events involving more than five people killed or wounded; events resulting from mishandling by railway workers or by failure, impairment, or breakage of carriages or railway equipment; events involving mainline travel delays of more than three hours; and other exceptional events. Regarding transport disorder, railway companies must also inform the District Transport Bureau of train service cancellations, passenger train delays of more than thirty minutes, and delays of more than one hour for non-passenger trains. As outlined here, even a passenger train delay of thirty minutes is recognized as an accident from a legal point of view in Japan's railway industry.

Some major railway companies also define events involving much smaller losses as accidents. For example, passenger train delays of more than ten minutes and damage amounting to more than 100,000 yen are classified as accidents according to some company regulations. Among transport industries, only the railway business regards delays of just ten minutes as accidents, investigates them, and considers measures to prevent them.

INTOLERANCE OF ACCIDENTS CAUSES ACCIDENTS

Human error and violations are the major human-related factors among the various causes of accidents (Figure 9.4). Although definitions of human error and violations vary among researchers, we can essentially define *human error* as incorrect actions executed under certain conditions by operators who are *unaware* that their actions are incorrect at the time they take place. We can similarly define *violations* as incorrect actions by operators who are *aware* that their actions are incorrect at the time they take place. Both errors and violations have in common that the operators are unaware that their actions will cause accidents and that they are able to execute correct actions in similar but more ordinary situations. The distinction is important, because if operators are aware that their actions will cause accidents, then these become criminal actions in the same way as terrorism. It should be noted that incorrect actions caused by an operator's lack of ability are not included in the definition of human error or violation.

The occurrence of human error has been explained as a lack of correct actions, or more likely as the unconscious execution of incorrect action (automatic processing). Reason (1990), a prominent psychologist who studies human error, illustrated this automaticity of action with the concepts of frequency gambling and similarity matching. The lack of actions, or the execution of incorrect actions, can happen when the operator has insufficient conscious control or when the operator does not direct enough attention to the correct action. This can happen when operators must consciously execute actions in response to certain situations or cues (see Figure 9.5) (Shigemori, Inoue, & Sawa, 2006). In general, this mechanism of human error through a lack of conscious control or attention has been explained by the notion of working memory (Norman, 1981; Reason, 1990). Working memory is an assumed brain function that people use when they visualize a certain procedure, memorize something temporarily, or perform mental arithmetic. It has restricted capacity in terms of the number of functions that can be performed at the same time.

The model derives two contributing factors of human error: insufficient attention and incorrect schema activated by context. The factors affect occurrence of human

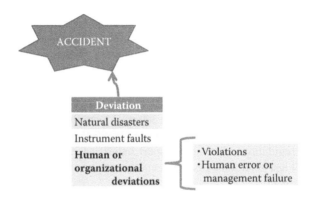

FIGURE 9.4   Primary causes of accidents.

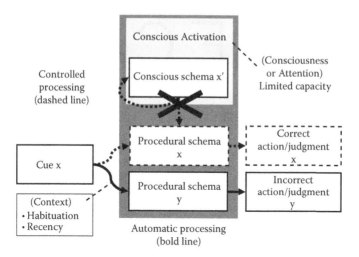

FIGURE 9.5 Model of occurrence of a human error. (Shigemori, Inoue, & Sawa, 2006. With permission.)

error in any stage of information processing, such as perceptual errors, judgment errors, and action slips. This fact has been confirmed by experiments (Shigemori, 2007a; Shigemori, Inoue, & Sawa, 2006). They used a digit-transcribing task in the input stage of information processing, the Luchins' water jar problem in the judgment stage, and a repeated letter-writing task in the output stage.

The digit-transcribing task required participants to transcribe handwritten percentage digits. The measuring standard for this task was set as the rate of incorrectly writing the target digit of "5%," which was mixed up with "50%" because the circle in the left-hand portion of the percentage sign was very large (Figure 9.6). The

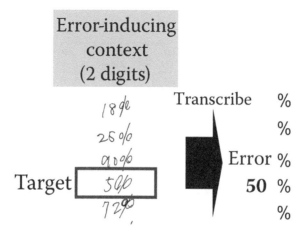

FIGURE 9.6 The digit-transcribing task in the input stage of information processing.

TΗΕ CΗT

FIGURE 9.7   Context effect on letter perception (Selfridge, 1955. With permission).

task also represented the classical illusion situation by context in cognitive psychology, involving the words "THE CAT" (Figure 9.7). Although the ambiguous symbol between "T" and "E" and between "C" and "T" are the same, we are prone to see the former as "H" and the latter as "A" (Selfridge, 1955). The task of reading the handwritten percentage terms was based on a near-accident in which a nurse misread a percentage term on a medical prescription written by a doctor (Shimamori, 2005).

The Luchins' water jar problem required participants to work out how to produce target amounts of water using three jars with different capacities (Luchins & Luchins, 1950). After repeatedly solving problems with similar patterns, participants become unable to solve target problems that had different answer patterns, even if they were easier than other problems (Figure 9.8). This task has been used to study functional fixation during problem solving. Such fixation occurs in the context of the repetition of similar answers, and it is difficult for those caught in functional fixation to find other answers. The phenomenon is similar to persistence in a decision-making situation, which induces errors in judgment.

The repeated letter-writing task required participants to write the Japanese phonetic symbol "□" (o) repeatedly. The exercise causes participants to write other characters incorrectly if they have similar stroke patterns, such as "□" (a) or "□" (mu) (Nihei, 1986a, 1986b, 1988) (Figure 9.9). Tasks involving incorrect writing or speech errors are frequently used to study action slips. In particular, repetition is familiar as a method for inducing slips.

In the experiments, the participants were divided into four groups by two conditions of attention (speedy or leisurely) and by two contexts (error-inducing context or

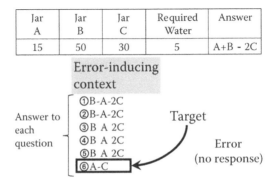

FIGURE 9.8   The Luchins' water jar problem in the judgement stage of information processing.

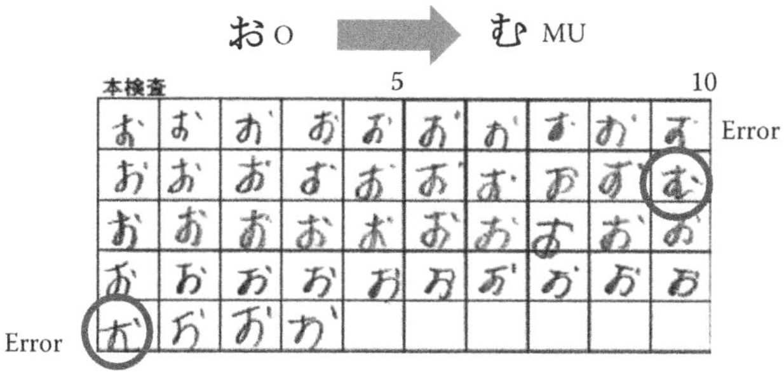

FIGURE 9.9 The repeating letter writing task in the output stage of information processing.

controlled context). In the speedy condition, the participants were asked to perform the letter-writing task as quickly as possible. The former condition corresponded to manipulation of the attention factor to correct action, and the latter corresponded to that of the context factor.

The results for all tasks showed that the error rates for each task grew in the error-inducing context and with the instruction to perform the tasks quickly (the inattention instruction) (Table 9.2). The results indicated that human error in any information processing stage can be caused by a lack of attention and/or context-induced activation of incorrect schema.

It has also been shown that 86% of assumed human errors with the potential to cause serious accidents, such as derailment or collisions, are due to the division of the attention needed for correct action (Table 9.3; Shigemori, 2003). In this study, 143 fatal human errors by train operators, assumed to be railway experts, were categorized by attentional factors (division or sustainment) and by factors of incorrect intrusion (habit, efficiency, antecedent, or omission). A fatal human error was defined as an action that leads to derailment or collision. It should be noted that most attention-diverting events were related to keeping schedules or preventing delays.

These studies suggest that one of the main causes of human error is the deviation of attention from correct actions and that the principal factor of attention deviation is the drivers' concentration on keeping schedules or preventing delays in the railway

TABLE 9.2
Percentage of Participants Who Made Errors on the Tasks

	Condition of Attention			
	Speedy		Leisurely	
	Context			
Task	Error-inducing	Controlled	Error-inducing	Controlled
Digit-transcribing	87.50	8.33	9.09	0.00
Water jar problem	62.07	0.00	26.07	7.14
Repeated letter-writing	55.56	13.3	30.77	36.36

TABLE 9.3

Ratios of the Causes of Assumed Human Error with the Potential to Cause Serious Accidents

Cause of Human Error		Ratio (%)
Lack of Attention	Type of Incorrect Action	
Division of attention capacity	Intrusion of habitual action	27
	Intrusion of efficient action	39
	Intrusion of action executed immediately before	0
	Lack of correct action	20
Attenuation of sustained attention	Intrusion of habitual action	2
	Intrusion of efficient action	10
	Intrusion of action executed immediately before	0
	Lack of correct action	2

Source: Modified from Shigemori (2003).

system. In conclusion, because the law and the railway companies consider a delay of a few minutes as an accident, and because railway companies strive to prevent such minor accidents, such obsessive attention to minor issues has the potential to cause other serious human error that possibly will lead to major accidents, such as derailment or collision.

BACKGROUND OF OBSESSIVE ATTENTION TO MINOR ACCIDENTS

There are two reasons why minor accidents are emphasized by the Ministry of Land, Infrastructure and Transport and by railway companies in Japan. One is the technology or culture of precise travel according to schedule, and the other is the false belief that major accidents can be prevented by eliminating minor ones.

Japanese railways have achieved unprecedented accuracy of operation in comparison to other transportation systems. It has been found that the mean delay time for Shinkansen trains is less than half a minute and that the corresponding figure for conventional railway lines in Japan is about a minute (Mito, 2001). Consequently, the activities of Japanese society depend heavily on the precision of the railway system. For example, businesspeople might make a minute-to-minute train schedule comprising a number of transfers in order to attend a meeting in Osaka, starting from their home in the suburbs of Tokyo. Most students also use railways more frequently than buses or taxis to attend entrance examinations. This dependence on a precise railway network causes people to seek even further precision in the system, and this excessive demand for accuracy reduces tolerance for train delays. In particular, people in the metropolitan area, who rely on dependable trains operating every few minutes, become impatient at even the shortest train delays.

The other reason for obsessive attention to minor accidents among railway company staff is the false belief in Heinrich's law and the broken window theory. First,

Heinrich's law states that for every major injury there are 330 accidents with no injury and 29 minor injuries (Heinrich, 1931). This thought pervades not only railway companies but also various industries and leads to the false belief that major accidents such as derailment or collision can be prevented by eliminating minor accidents or unsafe acts, such as delays of a few minutes or trains stopping at the wrong position in stations.

It is a frequent misinterpretation that risk management should concentrate on the control of unsafe acts (Hale & Bellamy, 2000). It is usually forgotten that Heinrich's data did not refer to the type of work situation typical of complex industrial operations today. In addition, although it is true that most serious accidents begin with unsafe acts, not all unsafe acts lead to serious accidents; there is a chain of several events between unsafe acts and serious accidents, so we can in principle control the risk of serious injury by breaking these chains, rather than by preventing the unsafe acts themselves. On the contrary, the obsessive attention to preventing minor accidents or delays of a few minutes has the potential to cause other human errors or major accidents. For example, it is considered that the 2005 derailment on the Fukuchiyama line in Japan may have been caused by the driver failing to control speed around a curve, because his attention was focused not on the operation of the train but on radio communication between the conductor and the director about a stopping error at the previous station (Aircraft and Railway Accidents Investigation Commission, 2007).

We must consider measures to prevent minor accidents caused by human error from growing into other human errors or major accidents, rather than focus on preventing minor accidents or human error per se. For example, railway companies must consider measures that will prevent deviations of train stopping positions in stations from growing into major accidents, by making drivers follow proper procedure after the mistake, rather than trying to prevent the deviation itself. Such minor accidents will never turn into major accidents if the driver does not behave improperly after the error, such as backing up the train without checking the position of the following vehicle. The current overemphasis on the prevention of deviation itself means that drivers pay no attention to proper procedure after mistakes, because of excessive concern about deviation; it may also cause them to take risks to catch up in the event of a delay.

Second, the broken window theory says that in order to prevent major crimes, we should not let minor offenses pass (Wilson & Kelling, 1982). Broken windows that lie neglected can be a sign that neighboring residents have no concern for problems such as window damage and that they may also lack concern with regard to criminal acts. It has been validated experimentally that broken parts on an abandoned car trigger destructive behavior (Zimbardo, 1969). Railway companies confuse this with Heinrich's law; but the broken window theory is about crime rather than accidents. Most felons commit crimes knowing that their actions are socially prohibited and illegal. In the case of human error, such actions represent violations, defined as actions that the operator knows are incorrect but which are not intended to cause damage to the system at that instant. Yet, violations should not be ignored even if they are easily excusable, because minor or forgivable violations that rarely cause accidents tend gradually to develop into major violations (Figure 9.10); a

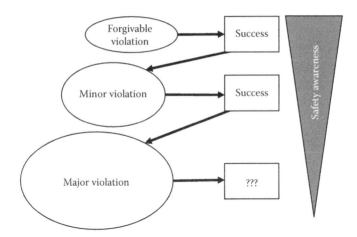

FIGURE 9.10 Development of forgivable violations into more serious violations.

few violations can lead to a deterioration of safety awareness and a working atmosphere that allows further transgressions by workers, causing violations to spread among staff.

In conclusion, we should apply Heinrich's law or the broken window theory to accidents caused by violations and should attempt to eliminate even minor transgressions. However, we should deal with accidents caused by human error with tolerance. The focus should also be placed on considering measures that can prevent human error, or minor accidents caused by human error, from leading to other human error, violations, or major accidents, rather than on striving to prevent human error or minor accidents caused by human error per se.

FUTURE ACCIDENT PREVENTION

Currently, Japanese railways recognize even a few minutes' delay as an accident and strive to eliminate such minor issues. However, such obsessive attention to minor accidents can lead to other human errors or violations that in turn may lead to major accidents. According to the broken window theory, to improve this situation, we should attempt to eliminate even minor (forgivable) violations. On the other hand, we should be more tolerant of human error that causes minor accidents and focus on measures to prevent major accidents derived from minor human error rather than on preventing minor human error per se.

To this end, we must change the way we think about human errors, and this applies not only to railways but also to people in general. If the general public has the attitude that railway companies should not allow even minor human error, such companies might not take a lenient attitude toward this matter in the face of such public opinion. Because it is difficult for railway companies to prompt the general public to be lenient toward minor human error, it is consequently up to academic experts to help build a correct public appreciation.

As a matter of fact, it is necessary to alter the perceptions of the railway industry as well as those of the general public. An effective approach is to establish a climate of informally discussing issues and of sharing stories of human error or minor accidents among workers. It is important for staff to routinely discuss and consider that all workers are constantly prone to human error, to consider the kinds of major accident that errors can cause, and to think about how to prevent such major accidents. This approach might make it possible for workers to raise safety awareness for post-accident procedures and to establish an atmosphere in which violations are not accepted.

Many years ago, workers regularly chatted informally in crew rooms or at lunchtime about human errors that they or other workers had made and about countermeasures they had devised for themselves. They shared information, not only on risks and countermeasures, but also on attitudes toward safety through such communication. However, these informal chats have now almost disappeared in a number of railway companies. This has happened for a number of reasons, such as the age gap among workers, reduced opportunities for such communication due to personnel cuts, changes in the working environments, and an increased number of people who do not want to relate to others.

To resolve the issue, some major railway companies support small group discussion activities for workers on accident causes and prevention. Currently, we are joining up with them and studying ways to share information on human error or hazards, countermeasures, and the experiences of individual workers through such activities (Shigemori, 2007b; Shigemori et al., 2007).

This kind of semiformal activity, involving the discussion of accident causes and prevention measures, is called accident roundtable discussion (ARD). The meetings consist of one or two facilitators and five to ten participants, who discuss the causes and prevention of a prepared accident for approximately ninety minutes. The most important point of the ARD system is that participants are encouraged to talk about their experiences in any area at all, not in order to find the easy answers, but to become aware of the difficulties of accident prevention.

The procedure consists of three primary stages: (1) discussion of accident image (DAI), (2) discussion of accident causation (DAC), and (3) discussion of accident prevention (DAP) (see Figure 9.11) (Shigemori & Miyachi, 2004; Railway Technical Research Institute, 2007).

In the DAI stage, participants discuss the process, the situation, and the seriousness of the accident. The aim of the DAI stage is to understand the scene and the seriousness of the accident sympathetically, as the subject under discussion. Those involved become capable of regarding the accident as an impending danger and participate in discussions in earnest. The DAI is subdivided into two parts: the first is visualizing the accident process and situation, and the second is quantifying the severity of the accident. For the first part, the facilitators show the participants a time serial collation table (TSCT) of the accident process (Table 9.4). The attendees then discuss the deviation events and similar experiences they have had. For the second part, participants discuss the worst possible scenario for the series of adverse events.

In the DAC stage, the participants discuss accident causation based on the repeated cause analysis method. The facilitators ask the participants why the deviation

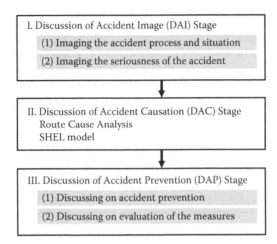

FIGURE 9.11   Procedure of accident round table discussion (ARD).

occurred and then ask why the cause occurred if a response involving a cause of deviation is given. Through this repeated question-and-answer process, the attendees discuss the causes of the accident in depth. If the discussion drifts from the main direction (e.g., it begins to involve private matters), the facilitators ask questions from a different viewpoint or from the view of another worker. We frequently use the SHEL model (Hawkins, 1987; see also Chapter 7) to prompt discussion from various viewpoints. The model offers various human-centered approaches based on human factors: liveware (humans), liveware-software (procedure), liveware-hardware (tools

TABLE 9.4
Example of a Time Serial Collation Table (TSCT)

Course of Events During Accident	Normal Course of Events	Outcome
The driver started the inspection before departure from the depot.	The driver starts the inspection before departure from the depot.	Deviation
The driver completed the inspection. The driver checked the time.	The driver checks the time.	
The driver waited at the driver's desk until the departure time.	The driver completes the inspection.	Deviation
	The driver waits at the driver's desk until the departure time.	
...	...	...
	The driver calls out the color of the signal for Track No. 1.	Deviation
The train ran beyond the go signal for Track No.1 even though it was red.		Accident

or machines), liveware-environment (work environment), liveware-liveware (relationships or communication). We also encourage participants to discuss their own experiences related to the causes.

In the DAP stage, participants discuss accident prevention. The aim of this stage is not only to consider measures against such accidents but also to provide awareness of the difficulty of accident prevention and to enable sharing of the participants' efforts. For this reason, the DAP is also subdivided into two parts: a discussion of accident prevention based on the participants' efforts and a discussion involving the evaluation of countermeasures. For the former, the facilitators encourage the attendees to talk about their efforts and measures to prevent accidents from two directions: self-performed countermeasures and countermeasures demanded by management. Most revision of rules or introduction of instruments as a means of accident prevention is attributable to management. Such countermeasures are important and efficient, but they do not enhance the safety awareness of the staff. To enhance safety awareness, self-performed countermeasures must also be considered. For the latter of the two DAP subdivisions, the facilitators encourage participants to evaluate the countermeasures devised through their own discussion, in terms of, for example, whether they are sure to prevent accidents, whether they can be kept up, or why they have not been implemented before. Such discussions increase participants' awareness of the difficulties of preventing accidents.

CONCLUSIONS

The prevention of serious accidents is a matter of priority in many industries. However, striving to eliminate all accidents can, rather paradoxically, have the opposite effect. Humans are not able to do everything and must therefore choose and concentrate on what needs to be done; that is, we must select the prevention of serious accidents and focus on it. To this end, we should be strict with violations but also have a more tolerant attitude toward human error.

The Japanese, however, have a very strict attitude toward human error rather than toward violations. This inverse approach by industry, the media, and the public is a factor behind misleading food labels on the one hand and derailment overturning accidents on the other, and has recently become a serious issue in Japan. It is, however, hard to change such attitudes, because they are underpinned by social and cultural factors.

The ARD system introduced in this chapter is a tool to change the attitudes of workers and managers toward accidents, violations, and human error. It is of course impossible to expect a dramatic shift as a result of the ARD, but a gradual shift is better in such complex social and cultural issues. The ARD is also just one of the tools available to help change attitudes. The application of various approaches, such as CRM training (International Civil Aviation Organization, 1989), an incident report system, and training on hazard perception (Crick & McKenna, 1991), is also believed to be useful. Furthermore, it is important to instill a similar view of accidents, violations, and human error in the general public.

REFERENCES

Aircraft and Railway Accidents Investigation Commission (2007). *Tetsudo jiko tyosa hokokusyo: Nishi-nihon ryokaku tetsudo kabusiki gaisya Fukuchiyama-sen Tsukaguti-eki —Amagasaki-eki ressya dassen jiko (Railway accident analysis report: Derailment between Tsukaguchi station and Amagasaki station on the Fukuchiyama line, West Japan Railway Company)*. RA2007-3-1 (in Japanese).

Central Japan Railway Company (2007). *JR Tokai kankyo houkokusyo (Central Japan Railway Company Environmental Report)*. Aichi, Japan: Central Japan Railway Company (in Japanese).

Crick, J. & McKenna, F. P. (1991). Hazard perception: Can it be trained? *Behavioural Research in Road Safety*, *2*, 100–107.

Hale, A. & Bellamy, L. (2000). Focused auditing of major hazard management systems. In I. Svedung & G. G. M. Cojazzi (Eds.), *Risk management and human reliability in social context* (pp. 23–41). Proceedings of the 18th ESReDA Seminar, European Commission, Joint Research Centre.

Hawkins, F. H. (1987). *Human factors in flight*. Hants, England: Gower Technical Press.

Heinrich, H. W. (1931). *Industrial accident prevention*. New York: McGraw-Hill.

International Civil Aviation Organization (ICAO) (1989). *Cockpit resource management (CRM) training*. Human Factor Digest No. 2, Flight crew training: Cockpit resource management (CRM) and line-oriented flight training (LOFT). ICAO circular, CIRCULAR 217-AN/132, 4–19.

Luchins, A. S. & Luchins, E. H. (1950). New experimental attempts at preventing mechanization in problem solving. *The Journal of General Psychology*, *42*, 279–297.

Ministry of Land, Infrastructure and Transportation (1987). *Regulation for railway accident report*. Departmental regulations, No. 8, Article 3.

Ministry of Land, Infrastructure, Transport and Tourism (2008). *Tokutei honpou jigyousya ni kakaru jouhou (Information related to specific airline companies in Japan)*. Ministry of Land, Infrastructure, Transport and Tourism, Civil Aviation Bureau (in Japanese). (http://www.mlit.go.jp/common/000016063.pdf)

Mito, Y. (2001). *Teikoku hassya: Nihonsyakai ni surikomareta tetsudou no rhythm (Departure on time: Railway rhythm in the Japanese society)*. Tokyo: Kotsushinbunsha (in Japanese).

Nihei, Y. (1986a). Experimentally induced slips of the pen. In H. S. R. Kao & R. Hoosain (Eds.), *Linguistics, psychology, and the Chinese language* (pp. 309–315). Hong Kong: University of Hong Kong.

Nihei, Y. (1986b). Dissociation of motor memory from phonetic memory: Its effects on slips of the pen. In H. S. R. Kao, G. P. van Galen & R. Hoosain (Eds.), *Graphonomics: Contemporary research in handwriting* (pp. 243–252). North-Holland: Elsevier.

Nihei, Y. (1988). Effects of pre-activation of motor memory for kanji and kana on slips of the pen: An experimental verification of the recency hypothesis for slips. *Tohoku Psychologica Folia*, *46*, 1–7.

Norman, D. A. (1981). Categorization of action slips. *Psychological Review*, *88*, 1–15.

Railway Technical Research Institute (2007). *Tetsudo Souken-shiki hyuman fakuta bunsekihou handobukku (Handbook of RTRI method of human factor analysis of accidents)*. Tokyo: Railway Technical Research Institute (in Japanese).

Reason, J. (1990). *Human error*. Cambridge, UK: Cambridge University Press.

Selfridge, O. G. (1955). *Pattern recognition and modern computers*. Proceedings of the Western Joint Computer Conference, New York, pp. 91–93.

Shigemori, M. (2003). *Classification of human errors by the mechanism of occurrence*. Proceedings of the 7th Joint Conference of Ergonomics Society of Korea/Japan Ergonomics Society (CD-ROM), Seoul.

Shigemori, M. (2007a). *Do human errors in different information processing stages occur by different mechanisms?* Paper presented at the 7th Biennial Meeting of the Society for Applied Research in Memory and Cognition, Lewiston, ME.

Shigemori, M. (2007b). Anzenisiki koujou wo mokuteki tosita jiko no group tougisyuhou (Group discussion method of accidents for safety-awareness building). *Railway Research Review, 64,* 22–25. Railway Technical Research Institute (in Japanese).

Shigemori, M., Inoue, T. & Sawa, M. (2006). Tasks for estimating human error tendency. *Quarterly Report of RTRI, 47,* 199–205.

Shigemori, M. & Miyachi, Y. (2004). *Tetsudo Souken-shiki hyuman fakuta jiko no bunseki-syuho (RTRI method of accident analysis).* Proceedings of the 12th Annual Symposium on Reliability, pp. 11–14 (in Japanese).

Shigemori, M., Suzuki, F., Aonuma, S. & Kusukami, K. (2007). Method of discussion on accident cause investigation in small groups. *The Japanese Journal of Ergonomics, 43,* 66–67 (in Japanese).

Shimamori, Y. (2005). *Iryoujikoboushi notameno hiyarihattojirei-tou no kijutujouhou no bunseki ni kansuru kenkyuuhoukokusyo (Report on analysis of incident report data for the prevention of medical accidents).* Multidisciplinary research enterprise for assessment of medical technology by grant-in-aid for scientific research of the Ministry of Health, Labour and Welfare in 2004 (in Japanese).

Wilson, J. Q. & Kelling, G. L. (1982). Broken windows: The police and neighborhood safety. *The Atlantic Monthly, 249,* 29–38.

Zimbardo, P. G. (1969). *The Human Choice: Individuation, Reason, and Order Versus Deindividuation, Impulse, and Chaos.* In W. T. Arnold & D. Levine (Eds.), *Nebraska symposium on motivation, 17,* 237–307. Lincoln, NE: University of Nebraska Press.

10 Organizational Learning for Nurturing Safety Culture in a Nuclear Power Plant

Hirokazu Fukui and Toshio Sugiman

CONTENTS

INTRODUCTION

The concept of organizational learning was originally proposed by Argyris (Argyris & Schön, 1978; Argyris, 1992) and has been practically argued by Senge (1990). Organizational learning has since attracted many researchers and practitioners who are interested in organizational changes, whether incremental or revolutionary. The concept represented an important function that had been overlooked as part of the managerial function taken as a major characteristic of organizations. Argyris argued that a wide range of organizational members required learning in order to revise their underlying thinking rather than just to deal with the challenges they faced. Senge proposed that a "dimension that distinguishes learning from more traditional organizations is the mastery of certain basic disciplines or 'component technologies'." He identified five disciplines that converged to "innovate learning organizations."

They were (1) systems thinking, (2) personal mastery, (3) mental models, (4) building shared vision, and (5) team learning.

Among the many studies concerning organizational learning, some focused on safety in nuclear power industries. For example, the EU project LearnSafe investigated facilitators and hindrances for organizational learning (Wahlström et al., 2005). In their project a series of empirical qualitative studies were conducted with the collaboration of senior managers of nuclear power plants in five different European countries. On the basis of empirical studies in nuclear power plants as well as chemical plants, Carroll, Rudolph, & Hatakenaka (2002) proposed a four-stage model of organizational learning that consisted of (1) local learning by decentralized individuals and work groups, (2) constrained learning in a context of compliance with rules, (3) open learning prompted by acknowledgment of doubt and desire to learn, and (4) deep learning based on skillful inquiry and systemic mental models.

The objective of the study reported in this chapter is to explore how safety culture can be nurtured in a nuclear power plant from the perspective of organizational learning. The study had two major characteristics: theoretical and empirical:

- We used Engeström's activity theory (1987) as the theoretical basis for the organizational learning in our study. Specifically, everyday organizational activities for organizational learning were conceptualized as a set of two interrelated kinds of activity: performance activities and improvement activities. From that set, existing everyday activities could be transformed into new ones by a kind of activity named transformational activity.
- During our five-year fieldwork in a nuclear power plant we collected specific instances of attempted methods that could contribute to enhancement of organizational learning for safety culture. Each example was positioned within the conceptual model outlined here. The model was developed gradually while carrying out the long-term field work, but in this chapter the model will be introduced first, followed by descriptions of specific instances, for the convenience of readers.

A CONCEPTUAL MODEL

ACTIVITY THEORY

Activity theory emphasizes the fundamental social nature of individual actions. In their everyday lives people tend to focus on an individual person when he or she shows either good or poor achievement. Good achievement tends to be attributed to the innate characteristics and attitude of the person, such as excellent ability or enthusiasm, whereas poor achievement tends to be attributed to lack of ability or lack of motivation. It is true that psychological factors such as ability and motivation sometimes are critical and thus should be targeted if an achievement has to be improved. However, a seemingly individual action often occurs as a part of the larger phenomenon of human collectivity.

Activity theory expands the field of view and thereby makes it possible to place what looks like an individual action in a larger context of collectivity. Let us start with an individual action and then expand our scope of view step-by-step by following

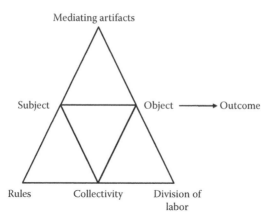

FIGURE 10.1 Structure of activity.

activity theory. At the beginning, an individual action is conceptualized in terms of a subject who works on an object and produces an outcome. For example, a particular worker, as a subject, who is responsible for the maintenance of a particular piece of equipment in a nuclear power plant, works on the equipment as an object, and produces it in good functional status as an outcome. Such an individual action is represented by a central subject–object→outcome horizontal line in Figure 10.1.

An individual action is always mediated by tools, in the sense that a subject works on an object to produce an outcome with the use of any available tool. Tools might be physical, such as a computer or an operation manual; institutional, such as an award system; linguistic, such as technical jargon; informational, such as specialized knowledge; or human, such as someone you can ask for minor technical help for your computer work. Importantly, tools are sustained and are made available for the individual agent by human collectivity. A computer on your desk is made available for you by the diverse efforts of many people who have worked from the manufacturing stage to the sales stage of the computer, and of many other people who have been involved in software development and Internet-related business. Even the person who is in charge of a mail server in your organization is part of the collectivity that makes the computer available as a tool for your electronic communication. Thus, using a tool always means putting yourself in collaborative relations with people who have made the tool available for you. In the terminology of activity theory, any action to work on an object to produce an outcome is always mediated by tools, or artifacts, that are sustained by a collectivity. In this sense, such tools are called mediating artifacts. These are shown in the uppermost triangle within Figure 10.1 (subject–mediating artifacts–object).

An individual action is carried out in more extensive collaborative relations with others than we see in mediating artifacts. That is true even if you are working alone at a particular point in time. You may bring documents that you have completed alone to someone else and ask for assistance to do some work using the documents. This process demonstrates that your work, writing the documents alone, is not a purely individual action, but that it is carried out as a part of collaborative work with someone else. Activity theory locates a seemingly individual action in the work of a collectivity (which was called a community by Engeström, 1987). This is shown in the lower middle small triangle in Figure 10.1 (subject–collectivity–object).

Having taken a collectivity into our scope of view, we can specify additional details of the collectivity in two ways. First, we can make clear what division of labor is maintained in the collectivity. The role of a subject in the division of labor is working on an object to produce an outcome, already represented by a central horizontal line in Figure 10.1. Then, we can clarify what role is played by each of the other members in the collectivity, shown in the lower right small triangle in Figure 10.1 (collectivity– division of labor–object). Second, it is useful to grasp what rules are shared in the collectively, explicitly (consciously) or implicitly (unconsciously). A rule concerns either fact recognition or value judgment. This is shown in the lower left small triangle in Figure 10.1 (subject–rules–collectivity).

We now have an entire structure of activity that consists of a total of six terms. The structure enables us to expand our scope of view so that an action, which at first glance was taken as an individual phenomenon, now can be located as a part of the larger collectivity. We have put an individual at the position of the subject in Figure 10.1 so far, but it is sometimes possible or even necessary to put a group of persons at the position of the subject, so that an action by the group can be understood as a part of the larger collectivity beyond the group.

From a practical viewpoint, a structural figure of activity provides more ways to improve the object→outcome relationship. While focusing on the horizontal central line in Figure 10.1, all you can do is improve an individual's ability, motivation, or personality. In many cases, however, it takes a great deal of energy and time to change the individual by education, training, or personal guidance, although those efforts sometimes should not be avoided. But there are many more options for improvement if you depend on a structure of activity. For example, you might want to improve the object→outcome term by introducing a new tool (a mediating artifact). You might also want to invite someone who can support the subject and thus create a new team (collectivity). Or you might want to change the role played by each member of the collectivity. Or you might want to change a shared belief (rule) in the collectivity by challenging a conventional way of thinking.

PERFORMANCE ACTIVITY AND IMPROVEMENT ACTIVITY

We will conceptualize organizational learning from the perspective of activity theory. Primarily, we should remember that organizational learning occurs in an organization and that the fundamental activities required of any organization (i.e., performance activities) therefore should be considered prior to learning. The most salient characteristic of an organization, as a kind of collectivity, is its artificial construction when compared with other kinds of collectivities such as a family or a community. Enrollment in and withdrawal from an organization is artificially determined by the discipline. Each organizational member is assigned a job responsibility according to vertical (hierarchical) and horizontal (lateral) divisions of labor.

In this sense, performance activity, in which a job assigned to each member is carried out, is indispensable for an organization to maintain itself as an organization.

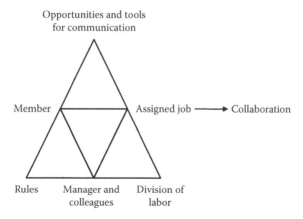

FIGURE 10.2 Performance activity for organizational learning.

However, just performing one's assigned job does not increase organizational learning. For organizational learning to be facilitated, performance of an individual member should be recognized by other members through communication, so that collaboration can be attained if necessary. The performance activity to facilitate organizational learning is shown in Figure 10.2, in which a member (subject) works on the assigned job (object) to produce collaboration (outcome) with the use of opportunities for communication, such as formal or informal meetings (mediating artifact), together with a manager, colleagues, or other people working in different workplaces (collectivity).

Now we can take a step forward to consider another kind of activity that has more to do with learning than performance activities; that is, improvement activities. The motivation to initiate organizational learning is small when organizational members discover what should be changed or improved in their workplace while carrying out their own jobs. However, it is difficult for them to keep remembering it, because they have been working in conventional activities that have been sustained by many people for a long time. Any crucial deficits have already been remedied; otherwise, the conventional activities would not have continued to work. By nature, conventional activities are those work habits that you feel comfortable doing, take for granted, and can rely on with ease. If you make small discoveries, conventional activities tend to make them temporary and then disappear, even though they could bring about valuable change.

Communication plays a critical role in sustaining small discoveries and in developing them into a possible way to improve workplace and work procedures. Your small discovery can get out of your interior world and become a topic of conversation if you can talk about it with someone else. Or, with the help of someone else, you can start an action to change a situation according to your discovery. Of course, it may not be easy to find such a person to talk to, because of the conventional activities in which you are embedded. Your discovery might be taken as something superfluous without which everything goes well, even if you do try to talk to someone about it.

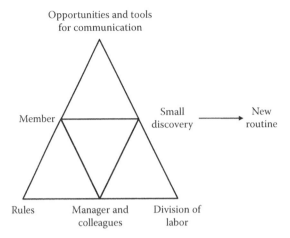

FIGURE 10.3 Improvement activity for organizational learning.

For organizational learning to take place, small discoveries should lead to an opportunity to change an existing routine into a new one through communication. The improvement activity for organizational learning is shown in Figure 10.3. For example, if you find that a particular maintenance procedure should be changed to check a particular portion of machinery more carefully by two different persons, it is not yet an improvement of activity. For an improvement of the activity to occur, you (subject) work on the small discovery (object) to produce a new routine (outcome) through a meeting and revision of a manual (mediating artifacts) with the collaboration of your manager and colleagues (collectivity).

With these two kinds of activity we can conceptualize everyday organizational activities for learning as the ones that consist of both performance activities and improvement activities. It should be noted here that the two kinds of activity influence each other in organizational learning. That is, collaborative practices attained in performance activities can expand the possibility of small discoveries in improvement activities. If you work closely with other persons, you can learn a new way of observation and thinking, which increases the possibility that you will find something new in your work. Also, a new routine developed by improvement activities may change the job assignment of each member in performance activities.

TRANSFORMATIONAL ACTIVITY

Engeström (1987) proposed a concept of learning activity, a kind of activity that works on an existing activity (object) to produce, or transform it into, a new activity (outcome), which stands on quite different premises from the previous one. That concept will be referred to as a transformational activity in this chapter. However, from our experience, to use the concept in a real setting, it is in many cases inconvenient to assume that a subject works directly on an existing activity and transforms it into a new activity. Rather, it is more natural to assume that an activity in which a subject

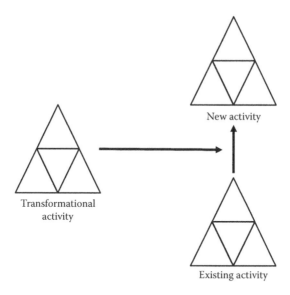

FIGURE 10.4 Transformational activity.

works on a concrete object and produces a concrete outcome, can lead to the transformation of an existing activity into a new activity. Taking this into consideration, transformational activity is defined as shown in Figure 10.4.

Organizational learning can be conceptualized comprehensively by combining a concept of transformational activity with the two kinds of activities discussed previously: performance and improvement activities. First, everyday organizational activities that facilitate organizational learning are constructed by a set of two interrelated kinds of activity, performance and improvement activities. Second, everyday organizational activities can be transformed by transformational activity into new ones, the premises of which were beyond imagination in previous everyday activities.

Transformational activity does not occur as often as improvement activity. But, two types of triggers can activate it. One is an extrinsic trigger in the sense that it is brought about by an abrupt challenge to what organizational members are doing, coming from the physical or the societal context. A typical example of a physical challenge is a big accident involving loss of life. Such an accident happens suddenly, like a bomb attack in a war. An example of societal challenge is a rapid change of policy from "for" to "against" nuclear power generation, for example, caused by a change of a national or local government.

It is important, however, that not all extrinsic challenges, physical or societal, become triggers of transformational activity. Often an organization will devote almost all of its financial and human resources to a short-sighted recovery, without paying attention to the initiation of a new effort to change the organization itself. Therefore, it is critical to determine if an extrinsic challenge can become a trigger that creates transformational activity.

The other type of activating event is an intrinsic trigger in the sense that it arises from what organizational members are doing. In this case, a seed of transformational activity is planted and grows in a corner of the organization. It is often so small that it is easy for the vast majority of organizational members to ignore it. Actually, we often hear of success stories in which a small group of persons shared a new idea, tried to materialize it, persuaded people around them gradually to accept it, and finally obtained support of top management. Thus organizational activities changed drastically. It is very probable, however, that such success stories represent just the tip of the iceberg. Many potential seeds of change are hampered by being either ignored or criticized by people around them. Thus, like extrinsic challenges, intrinsic small attempts are critical and can be utilized as triggers of transformational activity.

SPECIFIC INSTANCES

FIELDWORK

Specific instances of attempts seen as contributing to facilitating organizational learning for safety culture were observed in workplaces and collected in our field-work. The fieldwork was carried out in the maintenance departments of nuclear power plants at an electric power company in Japan. The company owns three sites, each of which has three or four units (a unit is a system including a single reactor). In each site, there are four departments responsible for maintenance of the primary systems (reactor), the secondary systems (turbine), the electric systems, and the measurement systems, respectively. Each department consists of two sections that differ in equipment and machinery, and each section has two or three work groups, each of which is composed of several workers and a head of the work group.

In the company under study, as well as in other electric power companies in Japan, physical labor tasks to directly inspect and repair equipment or machinery are implemented mainly by workers who belong to subcontracting companies. The electric company employees are involved mainly in preparing the necessary documents and making contracts for regular inspections (once every thirteen months) and daily maintenance. However, electric company workers are encouraged to visit the operation field where the equipment or machinery for which they are responsible is located and to maintain communication with subcontractor employees.

Our fieldwork was carried out in six maintenance departments at three company sites for five years. Interviews were conducted twice or more with all members in each department, including a department manager, section chiefs, deputy section chiefs, heads of work groups, and rank-and-file workers. A single interview took fifteen to thirty minutes. At the same time, we could always observe their workplace, because interviews were done at a table located in the corner of the workplace. We were often allowed to observe a regular meeting that was held in the workplace. Printed or written documents as well as information shown by a computer display were provided for us, unless it was inconvenient for the department.

RESULTS AND DISCUSSION

Instances of performance, improvement, and transformational activities that we found useful for organizational learning for safety culture will be described in the next few sections, following the structure of activity shown in Figures 10.2–10.4. In each instance, the most pertinent term among the six terms in the structure of activity shown in Figure 10.1 will be noted in brackets in the text; for example, [mediating artifacts].

Performance Activities

Job Performance on a Group Basis

You look at a work group in an organizational chart, but it makes a huge difference whether workers are expected to perform their jobs individually or in groups. In our fieldwork, the latter case, in which a work group functioned successfully as a group, was often observed. In those work groups, members as teammates gave advice and helped each other when someone was not confident about how his or her job should be performed or when someone could not complete a job alone [rules].

The head of a work group was a key person who had great influence on whether the work group could function effectively as a group. Most of the heads we met in our fieldwork had few machines for which they were directly responsible, or had only minor responsibilities for peripheral facilities. Instead, the head paid attention to the progress of each member's work and secured their performance by giving adequate advice, offering assistance, accompanying a member to an operation field, and assigning a role of assistance to someone else for a person whose work was temporarily beyond his or her capacity [division of labor].

A position as the head of a work group was assigned to an experienced person, without any necessary link to a route of promotion to a higher status. Therefore, it was not exceptional for the head of a work group to return as an ordinary group member after experiencing the position for several years. Effective functioning of the work groups as groups was sustained by leadership-by-experience rather than leadership-by-status.

In addition to the leadership by the head of the work group, we found other efforts that improved communication and collaboration. First, we found a group norm that facilitated as much help as possible for a person, when that person looked like he or she wanted it. One person said to us, "I am glad to help others if needed, and then I can enjoy being helped if I want" [rules].

Second, desks in the workplace were arranged in a way that emphasized fostering young or untrained workers. This was unlike the typical Japanese company, where desks are aligned according to status or age. Specifically, desks of young or untrained workers were arranged so that they were located between two experienced workers, which made it easier for the former to consult with the latter and for the latter to pay attention to the performance of the former [mediating artifacts].

Mutual Assistance Beyond Work Group

In one department we found that effective communication was maintained not only within a work group but between different work groups or sometimes beyond the department. In the department, workers who belonged to other work groups, and even other departments, were included in a collectivity in the structure of activity. Their role was to provide necessary assistance or give comments informally while being sustained by a rule that encouraged such mutual support [collectivity, rules, and division of labor].

Two types of departments that attained effective communication were observed in our fieldwork. The first type was a department in which it was easy for workers of different work groups to share technical knowledge and skill, such as a department for maintenance of electric systems or measurement systems. In those departments a worker could talk about what happened in other work groups with someone who belonged there. We found, however, that shared expertise alone did not necessarily guarantee good communication among different work groups. For this to happen, it was also necessary to have effective leadership by the department manager responsible for the entire department, an active attitude by the head of a work group toward facilitating communication with other work groups, or experienced workers who had developed a personal network with people in the other work groups and/or the other departments.

In the second type of department that achieved effective communication among different work groups, it was difficult for us to identify a specific reason for such good communication. For this type, all we can say is that it was because of an organizational climate or atmosphere shared by the entire site, including the department. Actually, many interviewees referred to this unique organizational characteristic of their site when they were asked to specify reasons for good communication beyond a work group or a department.

From our fieldwork we concluded that different sites were characterized by different cultures. This was often pointed out by interviewees who had experienced two or more sites in their career. They disclosed their impression of the characteristics of the sites they had worked for in terms of culture or atmosphere. It might be natural for different sites to have different cultures because the history, facility, and natural or social environment differ from one site to another. At the same time, however, top management and other personnel change every several years. If this phenomenon, which can be referred to best as culture or atmosphere, is maintained, it should be an important research topic to be addressed in the future.

Regular Meeting

In each department where our fieldwork was conducted, two meetings were held each day for all members — one when they started working in the morning and the other after lunch. In the morning meeting, a department manager informed all members of what they should be aware of and each work group transmitted various kinds of information to all members in the department. At the end of the meeting, one member read their goals for high quality and safety. Following that lead, the members read their goals together. In contrast, the afternoon meeting was held separately by each work group with the head of the work group.

Needless to say, a regular meeting plays an important role as a mediating artifact by which one's own work can be connected with that of others, so that a collaboration can be achieved. In our fieldwork, two instances were found as a way to enhance effective communication and collaboration. First, in one department, we observed that the worker who read the goals also made a brief speech concerning safety. He could speak about safety concerns, such as his experience in a formal setting, or a personal experience, in which he recognized the importance of being mindful of one's own safety. It might look trivial, but it was a precious opportunity for this worker to present his idea in front of all of his colleagues. Generally, rank-and-file workers are not good at presenting their own ideas in front of many people, and they tend to hesitate to do so unless they are encouraged by someone like their department manager. In this case, the department manager initiated the process in which workers' speech enabled them to play an active role in the meeting and increase involvement in their workplace. We assume that possibilities of effective communication and collaboration among workers would be increased if they had more involvement in their workplace.

Second, in one of the other departments, a thirty-minute meeting was held by persons in a supervisory status (namely, a department manager, two section chiefs, two deputy section chiefs, and the four heads of the workplace) just after the morning meeting for all members in the department. In the meeting, each head of the workplace reported a work plan for the day, which made it possible for all participants to share information on what work would be carried out and how it would be performed in their department. The department manager also transmitted important information obtained from a meeting that was held by top and middle management every day. The thirty-minute meeting was run in such a way that members could talk freely about whatever they wanted. The head brought the information shared in the meeting to each work group in the afternoon meeting. We hypothesize that a small meeting by supervisors, combined with a morning meeting by all members in the department and an afternoon meeting in each work group, would be more efficient in achieving effective communication in an entire department, rather than a meeting of all members in each section, because the latter was still too large for individuals to say something freely and to secure attendance of all members in the section.

Visits of Top Management to Workplaces

Top management tends to be physically and psychologically distant from rank-and-file workers. Communication between the two is usually indirect, being mediated by middle management such as a department manager. But we found in a certain site that an effort of top management personnel to put themselves in the workplace of rank-and-file workers could not only reduce the distance between them, but also impress the workers with the idea that the workplace was really a focus of attention for top management. In this site, one of the top managers visited the workplace almost every day. He talked frankly with a department manager or a section chief, and sometimes with workers. He brought a formal seal with him so that he was always ready to execute his approval immediately on a document submitted by workers, to show his involvement without bureaucratic delay (division of labor).

This instance illustrated the importance of top management showing their strong concern for the workplace of rank-and-file workers in a visible way. This led to workers' identification with their site and workplace, which in turn increased the possibilities for collaboration among them.

Extended Leave

Knowing a colleague's job makes it easier to help him or her; having experienced that person's work does so even more. In this sense, an extended leave affects more than the person who takes it. The leave brings about a situation in which a colleague has to assume the duties of another, in addition to his or her own. In other words, it becomes a situation where members are forced to acquaint themselves with a job that is usually performed by another person, thus collaboration would be made easier in the future [mediating artifacts].

In a certain department we observed, the department manager actively encouraged workers to take extended leave to facilitate the outcome described above. The manager also did not hesitate to take an extended leave.

Improvement Activities

Study Session

A study session or off-the-job training can be a beneficial opportunity to increase the knowledge and skill of workers in technical and/or institutional issues, if it is run effectively. It can contribute to helping a worker grow as a subject on the performance activities. At the same time, it can grow workers to the extent that they have more capabilities to discover something problematic in their workplace [subject].

We observed long-term efforts to run an educational program for workers in two different departments. In one department, a study session was held for an hour once a week, except during a regular inspection period. It was run under the leadership of an experienced worker, one of two deputy section chiefs, who was not in a managerial position. This brought about a sense of "our session" among workers and was referred to as the major reason for success by the department manager. In this department, an educational program that consisted of various kinds of seminars and training outside the site was established on an annual basis. Workers left their workplace to receive education and training according to the annual plan, in addition to a monthly study session in the department.

In the other department, a weekly study session that reviewed problem cases had been conducted for about ten years. The study session was suspended during the regular inspection period, as mentioned, but when our study was conducted it was held at least once every two weeks during the regular inspection period. Each study session focused on problem cases and discussed lessons learned from them. The duration of the session was kept within thirty minutes. According to the deputy section chief who was in charge of the session, session topics were chosen by utilizing a database called the nuclear information archives (NUCIA). This database, developed by an organization that consisted of many private sectors in the nuclear power industry in Japan, collects information about problem cases that occur in its member organizations and makes it public via the Internet. In this instance, the time restriction (as short as thirty minutes) and the use of a database covering abundant incidents were found to be the major reasons the sessions were maintained frequently and over a long time.

Comments Written by a Department Manager

Small discoveries by workers can be facilitated by their leaders. In one department, we observed such efforts by the department manager. In that department, subcontractors submitted a document to company workers to inform them of a work plan that would be implemented the next day. The workers reviewed it and filled in what they expected the subcontractors to do in order to pay attention to worker safety. In most departments, the document was returned to the subcontractors after review, but in the department we observed, the manager further reviewed the document, added his comments by handwriting when necessary, and then returned it to the workers. It was time consuming for the department manager because it took two or three hours to review as many as twenty or thirty documents carefully, especially during a regular inspection period in which the number of documents was greater than usual. Doing so, however, allowed this department manager to teach his workers many things from his years of expertise in the area covered by the department [subject]. Also, the comments by the department manager could expand the workers' scope to look at their work and thus increase the possibility of finding something to be improved [mediating artifacts].

Revision of Checklist and Skill Transfer System

In almost all the departments we observed, various checklists had been developed to find omissions and mistakes by workers and subcontractors. However, although some checklists were utilized effectively and were revised following situational changes, others were used less and less frequently and were eventually ignored. Continuous revision of a checklist required attention to the checklist, combined with examination of one's work procedures. This indicated that revision of the checklist was useful in enhancing small discoveries by workers [mediating artifacts].

In our fieldwork, we were able to look at several checklists that were in use while being revised. There also may have been checklists that were once developed but had stopped functioning.

More sophisticated than a checklist was a database, which workers in a certain department developed to document a cumulative record of the skills and knowledge they obtained in their performance process. The database was called the skill transfer system. More than thirty skills and associated knowledge were input each year and more than three hundred items were accumulated. There was one person who contributed ten items in a single year. Another worker presented his input in a meeting to make his colleagues aware of it. Such a database was used more effectively by workers than other databases that were developed exclusively by technical staff people in the site and unilaterally given to workers, who were then forced to use them. An attempt to develop an intrinsic database is not possible without efforts to accumulate small discoveries in the workplace. However, it can also be useful as a mediating artifact by which small discoveries are facilitated.

Award System

We observed the implementation of an award policy in one site. Specifically, the number of suggestions for improvement proposed by workers and the number of qualifications obtained by workers were recorded and transformed into scores. Then, the department with the highest overall rank was given an award in a ceremony in which all workers in the site participated. The policy was utilized very differently from one department to another.

In one department, the department manager had declared his intent that workers try to obtain as many scores as possible in order to earn awards, encouraging them to make greater efforts for this purpose. He understood that the department was responsible for important but thankless jobs, and he was aware that it would be difficult to be appreciated by top management and other departments in the site, no matter how hard workers tried to do their jobs. In a situation like this, he thought, workers could show their dedication and visibility in the site and thereby increase pride in their jobs by being validated with awards. It is plausible that increased pride leads to more interest in a job and workplace, and thus facilitates greater opportunities to make small discoveries that might bring about improvements [mediating artifacts].

Transformational Activities

Generally, an instance of transformational activity can be observed much less frequently than performance and improvement activities for organizational learning. Fortunately, we were faced with what we may call transformational activity in our fieldwork.

In our interviews during these two years, most people mentioned that improvement requests made by subcontractors had drastically increased and that almost all such requests were being accepted and materialized through a budgetary adjustment by the company. As mentioned earlier, physical labor tasks in nuclear power plants in Japan are generally conducted by subcontractors' employees in both regular inspections and usual maintenance. Improvement requests made by workers of subcontractors are therefore important sources of information for improving the reliability of the plants as well as the safety of the workers. However, although a system for making improvement requests had existed previously, the number of such requests made by subcontractors was much smaller until a few years ago.

Such a system to materialize improvement requests by subcontractors will be referred to here as the materialization system of improvement requests, or MSIR for short. Although each of the three routes for the submission of improvement requests (described in the following) had a name, the system as a whole was not given any name in the company. In the MSIR, improvement requests of subcontractors were submitted to the company through one of the three routes. First, subcontractors and staffs of the company assessed possible risks in the workplace where subcontractors' employees were going to work, prior to the commencement of each project. If any possible risks were detected, corrective measures were taken. Second, after the work was completed during a regular inspection period, subcontractors submitted a work report to the company. The report included a portion where they could note whatever should be improved in machinery they operated and in the maintenance procedures. Those points were discussed in a meeting, in which heads of subcontractors and workers responsible for the machinery in the company participated, to clarify what should be improved and the details necessary to make it happen. Third, subcontractors could submit an improvement request form that was used among workers of the company. The company sorted these forms by content and assigned them to pertinent sections where their validity was deliberated.

According to activity theory, an activity following the MSIR is one in which subcontractors (subject) work on their present workplace (object) to produce an improved workplace environment (outcome) with the use of the MSIR (mediating artifacts) in collaboration with people of the site (collectivity). It had been almost three years since the MSIR was introduced in each site of the company. Workers of the company, as well as subcontractor employees, were accustomed to the MSIR, in which subcontractors could submit improvement requests without hesitation and almost all of them were implemented. Such a situation had already become common in each site.

The MSIR was started because several subcontracted employees, preparing a regular inspection several years ago, were killed by high-temperature steam that spewed from an exploded pipe in one plant of the company (see Chapter 1). In fact, it was the first fatal accident in the history of a nuclear power plant in Japan. Soon after the accident, inspectors discovered that the portion of the exploded pipe had not been inspected for more than thirty years. This accident was so huge that the company had to reconsider the organizational activities of the nuclear power plant from various aspects. One lesson learned from the accident was to pay much more attention to the operational field where machinery was working; another was to listen to information from people working there, especially information from subcontractors. The MSIR was initiated as a major approach to actualize the lesson. One of the most important members in the collectivity, in the activity following the MSIR, was in the top management of the entire company. He started and continued to promote the MSIR to the extent that almost all requests from subcontractors were honored.

The activity certainly transformed the underlying premises of organizational activities prior to the accident into new ones, based on a new premise concerning the necessity to pay attention to information from subcontractors and improve the operational field. In the new premise, the operational field was the focus of attention to an extent that had been beyond imagination until the accident. In this sense, an activity based on the MSIR can be referred to as a transformational activity that was initiated with the accident as a trigger.

In addition to the MSIR, we had an opportunity to observe another instance of transformational activity in one of the other electric power companies in Japan. The company experienced a big problem several years ago, when it was accused of not reporting the discovery of a crack in the shroud of a reactor to the national governmental agency. Having learned a lesson from this bitter experience, the company established a special committee on which top and middle management were included to work beside persons responsible for problems and safety. The committee met for two or three hours each day, examining reports submitted from all over the site concerning any defects of facilities and compromised work conditions. For each reported case, countermeasures were planned along with a ranking of their urgency. A website that disclosed and posted the results of those examinations was an important improvement.

In this activity, the committee (subject) worked on a reported problem (object) to produce disclosed countermeasures (outcomes) with the use of the companywide system in which every deficiency and problem should be reported to the committee (mediating artifacts) with the collaboration of top management, company members of the site, and subcontracted employees (collectivity). The activity transformed the underlying premises, not only in the resolution but also in the informational disclosure of deficits, into ideas that had not been imaginable before the committee was started. The activity exemplified transformational activity that was triggered by an extrinsic event; namely, an accusation due to the omission of a report.

CONCLUDING REMARKS

In this study, organizational learning was conceptualized from the perspective of Engeström's activity theory and a conceptual model of organizational learning was proposed. Also, we collected specific incidents of attempts to contribute to facilitation of organizational learning for safety culture in a nuclear power plant and located them in the conceptual model.

At the end of this study, it is possible to point out the practical value for an organization in which our fieldwork was implemented. This chapter focused only on positive discoveries in our fieldwork, because the purpose was to collect instances that could contribute to enhancement of organizational learning for safety culture, as conceptualized in Figure 10.5. However, many interviewees in our fieldwork remarked on negative aspects. For example, during the first two years of our five-year study, almost all interviewees complained about a new online system that was introduced for comprehensive management of maintenance work, and a new inspection rule that was issued by the governmental agency.

We wrote a research report every year and submitted it to the top management of the site. The report included a summary of interviews and observations, conceptual or theoretical interpretation, and a list of all remarks, positive or negative, made by interviewees. We were careful to maintain confidentiality in composing those lists. Generally, it is difficult to maintain good communication flow from the bottom to the top in an organization that relies on a hierarchical structure for its function. This is because, basically, it is difficult for rank-and-file workers to say something freely, especially something negative, to top management even if they want to. That said, it might be useful to secure a route of communication flow by inviting researchers,

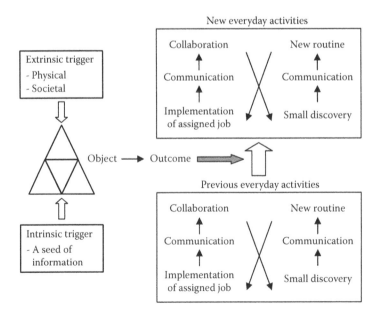

FIGURE 10.5 Comprehensive conceptual model of organizational learning.

such as the authors of this chapter, to listen to voices from the bottom of the organization and utilize them to develop a policy for top management.

In fact, this study was utilized by the site where we carried out our fieldwork. We prepared a two-page report to all workers in the department where our fieldwork was conducted, restricting its contents to positive examples for organizational learning. The report was not only for appreciation of their collaboration in our fieldwork, but also for providing materials to enable them consciously to understand their strengths. A situation in which one consciously does something good is different from a situation where one does the same thing without knowing it is good. Through our study we hoped that the employees came to share a vision and recognized their strengths for further enhancement. The president of the site distributed the short report in a meeting of top and middle management, to present it as an example of good practice from which managers might be able to learn something to improve their workplaces.

We will continue to collect specific instances in our future empirical work in a nuclear power plant. Concomitant to the fieldwork, we will develop our conceptual model further by taking into consideration actor network theory (e.g., Callon, 1987), core-task analysis (e.g., Norros, 2004), and situated learning theory (e.g., Lave & Wenger, 1991).

REFERENCES

Argyris, C. (1992). *On organizational learning* (2nd ed.). Oxford: Blackwell.

Argyris, C. & Schön, D. A. (1978). *Organizational learning: A theory of action perspective.* Boston, MA: Addison-Wesley.

Callon, M. (1987). Society in the making: The study of technology as a tool for sociological analysis. In B. Wieber et al. (Eds.), *The social construction of technical systems: New directions in the sociology and history of technology* (pp. 83–103). London: MIT Press.

Carroll, J. S., Rudolph, J. W. & Hatakenaka, S. (2002). *Organizational learning from experience in high-hazard industries: Problem investigation as off-line reflective practice.* MIT Sloan School of Management (Working Paper 4359-02).

Engeström, Y. (1987). *Learning by expanding: An activity-theoretical approach to development research.* Helsinki: Orienta-Konsultit.

Lave, J. & Wenger, E. (1991). *Situated learning: Legitimate peripheral participation.* Cambridge, UK: Cambridge University Press.

Norros, L. (2004). *Acting under uncertainty: The core-task analysis in ecological study of work.* Helsinki: VTT Publications.

Senge, P. M. (1990). *The fifth discipline: The art & practice of the learning organization.* New York: Doubleday.

Wahlström, B., Kettunen, J., Reiman, T., Wilpert, B., Maimer, H., Jung, J., Cox, S., Jones, B., Sola, R., Prieto, J. M., Arias, R. M. & Rollenhagen, C. (2005). *LearnSafe: Learning organisations for nuclear safety.* Helsinki: VTT Publications.

11 Establishing an Organizational Safety Culture

Michio Yoshida

CONTENTS

SAFETY AND GROUP DYNAMICS IN JAPAN

Group dynamics focuses on human behavior in the group and is one of the most important research fields in social science. The study of group dynamics was started by Kurt Lewin in the United States in the 1930s. Since then, many scientific analyses of the relationships between groups and social conflicts, personal relationships, leadership, etc., have been done as various problems in organizations and groups have occurred. Practical research conducted in real society has been integral to group dynamics.

In Japan, research on group dynamics started at Kyushu University in 1961. One of the prominent leaders was Juji Misumi, who conducted research in the area of school education at the first stage. After several years, he and his colleagues also focused

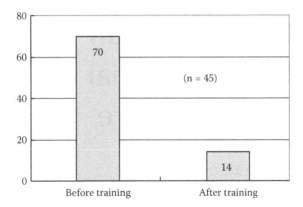

FIGURE 11.1 Number of accidents before and after the introduction of group decision-making methods.

their research on industries, which had been recovering from serious damage caused by World War II.

They started this line of research by looking at how to prevent accidents in a bus company where there had been a serious increase in the number of accidents (Misumi & Shinohara, 1967). Misumi and his colleagues introduced the concept of "group decision making," which Lewin had demonstrated to be effective in changing people's beliefs and behavior.

At the bus company, forty-five drivers who had been involved in more than two accidents each were gathered to participate at the training center. In the training, the participants discussed the problems in their workplaces and analyzed their behavior to realize safety in driving. After careful analysis of the accidents in which they had been involved, they ended their training by making a decision on how to avoid accidents in the future. As shown in Figure 11.1, the number of accidents drastically decreased after the training, compared with the number of accidents ten months before. These results clearly demonstrated the effect of group decision making.

After the success with accident reduction at the bus company, a shipyard in Nagasaki embarked on a safety campaign. The shipyard had seen an increasing production of large ships, mainly oil tankers, but also suffered from an increase in the number of accidents. The shipyard decided to introduce a new safety campaign to the workers using techniques mainly dependent on group dynamics. After two years of action research, the rate of accidents was drastically reduced. They were also able to get the workers more motivated, resulting in reduced costs and a high commitment to their products. Absenteeism decreased, possibly because family members were invited to the workplace, something which had not been expected at the start of the campaign (Koh, 1972).

This research led to a collaboration with the Institute for Nuclear Safety System (INSS), which was founded in 1992 by Kansai Electric Power Company in Japan. Misumi and his colleagues developed a scale for measuring the leadership of supervisors working at nuclear power stations and the safety climate in work units. They also developed new types of leadership training, which contributed to the improvement of interpersonal skills of supervisors as well as their followers.

The collaboration with INSS and JIGD, the Japan Institute for Group Dynamics, is still ongoing and has succeeded in developing a deeper understanding of nuclear safety.

KEY TERMS RELATING TO SAFETY CLIMATE

Almost all organizations try hard to improve the safety of the work environment, especially the safety of their tools and technology. But accidents, unfortunately, continue to occur everywhere in the world. One of the most important problems is that accidents cannot be eliminated completely only by making changes to the working environment. Human-related issues also need to be considered in order to avoid or prevent accidents and to establish a safety culture. The safety culture is in itself a problem involving groups of people. From this viewpoint, the theory and research experiences in the area of group dynamics should contribute to establishing the safety culture.

Here we would like to highlight some key terms and group norms that can be expected to be shared by members in workplaces.

FAIL-SAFE AND FEEL-UNSAFE

The concept of *fail-safe* is that if something fails, then there must be a safe solution that will counteract the failure. People working with technology should therefore always act based on this concept. *Feel-unsafe* means that if you feel that "something is wrong," then you should think "there may be a problem in safety."

The fail-safe concept is widely accepted as a basic concept for achieving safety. The idea is that if something goes wrong, the system tends to select the safest response. Almost all technology introduced in safety-conscious organizations is established as fail-safe. But even under fail-safe conditions, accidents continue to occur in various situations and their main causes are human error. From the viewpoint of group dynamics, we therefore focus on decision making, personal relationships, leadership, and communication in the organization, all of which concern the human side of an organization. There is a need to have a heightened sense of the inherent risk in situations when workers think that something is not quite right. And we must learn not to think of this sense of danger as merely a personal fault or misunderstanding. People tend to refrain from stopping machines because of the heavy cost of recovering lost production. This kind of feeling is easy to understand, especially from the supervisors' perspective. But it is also necessary that they feel unsafe in these conditions.

We would like to emphasize the importance of workers having the idea and sensitivity toward the concept of feeling unsafe. Technology should first of all be fail-safe, but humans should have the feel-unsafe attitude at the same time. The concept of feel-unsafe is extremely important for sensing risk. It goes without saying that people should make choices based on the situation that lead to a more secure outcome, or are fail-safe. The concept of feel-unsafe will be useful in these situations, because technology, such as machinery, cannot feel unsafe. In contrast, humans have the power to feel unsafe by utilizing knowledge and experience gained over time. This can, of course, also include intuition. It is in principle conceivable that machines will have this sort of ability at some point in the future. However, even then, the ability of humans to feel unsafe will continue to play an important role for the realization of safety. It is thus desirable to adopt fail-safe for design and operation of equipment or devices and feel-unsafe for the people who use such equipment or devices.

FROM KNOWLEDGE TO AWARENESS, THEN TO ACTION

Safety cannot be secured without *knowledge* about safety. But even with knowledge, accidents may still occur. Everybody knows that drunk driving is dangerous, but this knowledge is not always realized in people's behavior and unfortunate traffic accidents continue to happen. Thus, *awareness* is necessary to make use of the knowledge for safety. But just having awareness is not enough. Safety can be realized only when the awareness leads to action. Knowledge can be taught, but awareness cannot be taught. This viewpoint — from knowledge to awareness, then to action — is one of the key concepts for establishing a safety culture. It is related to human relationships, a sense of responsibility for jobs, pride, etc.

Methods for translating knowledge into action have been demonstrated through group decision making in the field of group dynamics. Lewin first demonstrated group decision making when trying to change eating habits (Lewin, 1947). Misumi and his colleagues have applied this method to accident prevention in Japan and have obtained very successful results. Lewin suggested three stages for behavioral changes: unfreeze–change–refreeze. The method of group decision making promotes this process effectively. It is also clear that the following are important for decision making: namely, satisfaction must be gained through participation in making one's own decisions, having open discussions, and the creation of group norms.

EVALUATION OF EMERGENCY RESPONSES

Evaluation of emergency responses is also important to achieve safety culture. When a typhoon is expected, various countermeasures are taken to minimize the damage. Some important events even may have to be canceled. But when the typhoon actually comes, it is often much weaker than people expected, causing almost no damage. Facing such a case, we tend to think that we should not have canceled the event or that we may have overprepared. But risk management should not be judged only by the result. First of all, we should be happy that nothing serious happened. Also, we should not regret having taken safety measures. It is important to believe in the correctness of our own decisions. Otherwise, the next time we face a similar situation, we may make an unsafe and risky choice. We should not feel regret and think, for example, "I did not need to take out life insurance" after having lived a long life. Such an attitude is against the spirit of risk management. There is always pressure toward cost reduction in organizations. The stronger it is, especially when safety measures are not connected with accidents, the more careless attitudes tend to become. The leadership of the top management has an important influence in such situations. The top leaders' decisions with regard to the spending of money to ensure safety will, in and of themselves, raise the safety awareness of the organization's members. When this has been established, a bottom-up viewpoint, which takes into account the opinions and ideas of the workers, is indispensable.

CERTAINTY RATHER THAN PROBABILITY

Equipment and machines are made based on the concept of fail-safe. Therefore, the *probability* of the occurrence of an accident is quite low, even if the rules listed in the procedures are not followed. In fact, a highly dangerous activity in the workplace may not immediately lead to an accident. But the development is just a matter of probability. When making choices concerning organizational safety or human lives, it is important for us to select *certainly* safe options rather than *probably* safe ones. Such choices are not just scientifically appropriate but are deeply related to the nature of the human mind. We tend to engage in risky behaviors when the probability of having accidents is low. The reason for this may be to achieve cost reduction or perhaps to save time. However, these actions will increase the probability of having an accident. It is important that all the members of the organization have a shared sense of values and awareness. We can expect safer actions to be taken if the norm "definitely safe" rather than "probably safe" is established in workplaces. Safety is therefore very much a matter of awareness of values, norms, and morals in society.

THE "IT HAS NEVER HAPPENED" SYNDROME

One excuse people often give after an accident is that "we did not think our actions would cause an accident because we had done it safely before." Strictly speaking, they may not have been completely certain about all of the safety implications. But they still thought that for some reason it would not lead to an accident.

On the other hand, there also exists the "it has happened before" syndrome. If a sign of trouble is found, no action may be taken because people think "a similar thing has happened before but it never caused a big problem," and thus they miss a chance for preventing future accidents. There is also a risk that such a trouble is regarded as something that "often happens," even though it may have actually happened "only once."

No activity is absolutely safe from accidents. The probability is never zero. Therefore, the formula *it has never happened = it will never happen* does not work. We need to keep in mind that perhaps it has never happened just because we were lucky. No one would say, "I will never have cancer" because "I have never had cancer before."

FOCUSING THE GROUP ASPECTS OF ORGANIZATIONAL SAFETY

SAFETY AND LEADERSHIP

We have conducted studies on leadership as it relates to morale and safety. In these studies, we found a consistent relationship between the leadership of supervisors and the morale of their subordinates. We also found that leadership affects safety behavior in workplaces. In these contexts, we have developed training programs to improve leadership and interpersonal skills. As for leadership, Misumi and his colleagues have developed the concept of the performance–maintenance, or PM, relationship (Misumi, 1985). In their field research, they found that there are two important dimensions to leader behavior: performance and maintenance. According to the degree of the two behaviors, leaders are classified as one of the following four types:

1. *PM*: A leader who does well with respect to both performance and maintenance behaviors
2. *P*: A leader who does well with respect to performance but is weak in maintenance behavior
3. *M:* A leader who does well with respect to maintenance but is weak in performance behavior
4. *pm:* A leader who is weak in both performance and maintenance behaviors

Many studies support PM-type leaders as having the best results for a groups' morale and productivity, as well as for preventing accidents (Misumi, 2001). The M-type comes in second followed by the P-type. The pm-type comes in last in almost all cases. Figure 11.2 shows the effect of PM leadership types on subordinates' work motivation. We also found important relationships between leadership and accidents. Figure 11.3 shows the relationship between the PM classification of the managers and the number of accidents in a bus company. Under a relatively high-performance and high-maintenance leader, the PM type, no accidents occurred. On the other hand, having a low-performance and low-maintenance leader correlated with driver accidents. The combination of leaders' high performance and maintenance behaviors seems to have a synergistic influence on their subordinates' motivation and satisfaction with regards to working situations, mental health, etc. This leads to an increase in the safety norm of the members.

We have made a scale for measuring P and M behaviors in a nuclear power plant as part of a joint research project between the JIGD and the INSS. A safety climate

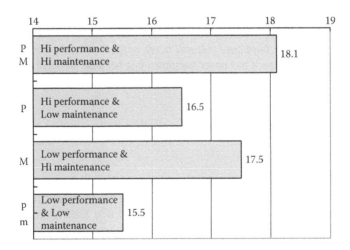

FIGURE 11.2   Scores of subordinates' work motivation under four types of leadership (5 items with 5-point scale varying from 5 to 25 points).

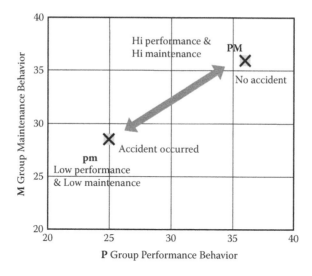

FIGURE 11.3   The relationship between accident occurrence and leadership of a bus company.

scale for a nuclear power plant has also been developed through this research. The main results were identical with those obtained in former studies. In many cases, the PM type was the most effective and the results of the pm were the lowest. Although the development of technology contributes to the prevention of accidents, we should also focus on the importance of leadership and interpersonal relations in workplaces. In this context, Yoshida (1999) has been developing and implementing training aimed at improving a supervisor's leadership and has found there are effects on subordinates' morale and other related aspects.

SAFETY AND THE HUMAN SIDE OF ORGANIZATIONS

Some accidents can be prevented by controlling the physical environment. For instance, railway crossing accidents do not happen with the Shinkansen trains, which run on elevated exclusive tracks. However, you should think of this as an exception. It is impossible to avoid completely mistakes and accidents, even though we try to improve and maintain technology. Sometimes limited express trains, highly equipped with computers, do not stop at stations where the display in front of the drivers tells them to do so. Sometimes airplanes, which the manufacturing companies tout as safe, crash. It is clear that such mistakes and accidents in many cases originate with people. This is the so-called human error. In a broad sense, all errors are ultimately human errors, even if they are due to design flaws in machinery. However, here we define mistakes as being human errors when they occur even if the technology, the procedures, and the rules are perfect. These errors are strongly affected by human nature and by interpersonal relations. Therefore, it is necessary to understand human characteristics sufficiently well to ensure the safety of organizations.

It is also important to find the universal causes that lie behind accidents. There are many commonalities in the causes of accidents. Case studies of accidents typically find workers who say, "we can't speak freely" or "safety problems expressed by members are not accepted by work groups."

The natural abilities that humans need in order to thrive can also be the very causes of mistakes and accidents. It becomes an important factor when such abilities cause lapses in judgment that may lead to accidents. It is ironic that our inborn habit-forming nature, which is such an aid to our survival, can also be the cause of our demise. Obviously, it is not easy to overcome this. Like an insidious devil, these habits take over our minds. We can express this by referring to the "devil's laws."

The Devil's Laws

The Devil Creeping in on the Requirements for Better Living

The first "devil" is getting accustomed (to something). In order to function smoothly, it is indispensable for us humans to get accustomed to actions and be able to perform them automatically or unconsciously. As newborns, we are hardly capable of doing anything. We must make efforts to do everything consciously. Through this process, by learning and training, we gradually get accustomed to doing certain things and begin to adopt unconscious behavior and habits. The ease of our daily life is assured by this ability, especially by getting accustomed. With this ability, we can drive better, do our jobs well, move up in the world, maybe eventually make speeches before audiences, and so forth. But at some point, the more accustomed we become to something, the less attention we tend to pay to our activity. This reduces our carefulness, and we start to see the daily work situation as "normal." At this stage, our unconsciousness reduces our sensitivity to unsafe situations. It is hard to go back to the basics and take a fresh look at things. We need to start thinking, "I'm doing well with my job because I'm accustomed to it. But this confidence could be dangerous for me. I should be more careful in doing my daily job." Very simple reflection of this kind can help exorcise the first devil.

The second devil's law is the wrong evaluation of past experience. An example is to think that things are safe because the accident has not happened up to now. This reference to the past causes one to tend to misjudge the present situation as being the same. Nobody thinks, "I will never get cancer in the future because I haven't gotten cancer yet." The fact of not having any accidents could in and of itself lead to a higher probability of a future accident. This might be seen as an example of the so-called gambler's fallacy. However, when it comes to safety as it relates to environmental deterioration and so forth, people have a tendency to try to view the probability as lower than it actually is, even when the probability of the occurrence of problems actually increases. In group situations, there is furthermore a strong force leading the members toward conformity. Rather than focusing on the probability of safety as a win-or-lose situation, it is necessary that judgments and decisions be made with the aim of greater safety. This notion needs to be embedded in the workers' consciousness in order to ensure workplace safety.

We also tend to embellish or fabricate experiences we have had. When an unusual thing or event happens, we may think "the same thing happened before." We might try to rationalize our behavior, especially if the situation is difficult. Even if the exception had happened only one time, the fact that it had happened at all would have a strong effect on our minds. Then rationalization follows, in the form of "we surely have experienced it before, so it will work out well this time too." In a group situation, this kind of rationalization is strengthened. The group members may therefore go on to the judgment that the thing is not an unusual occurrence. This is a kind of storytelling rather than a rationalization. The following example illustrates that:

> A mistake involving a patient undergoing a heart operation occurred at a hospital. The patient having the heart operation was not the right person, but was rather someone who had lung problems. The electrocardiogram had indicated a normal heart state before the operation. This should have been a strong indication that something was amiss and should have raised doubts as to whether it was the correct patient or not. The group of doctors, however, made their judgments based on their past experiences with anesthetized patients. One doctor said, "I've experienced the same case before." "Yes, I have also experienced it," followed another. In this situation, they did not evaluate the objective probability of the phenomena. Thus, the entire group began to try to convince themselves, and through storymaking this situation was rationalized and accepted by the group. The syndrome of "the same thing occurred before" had a fatal effect in this case.

Such a tendency also occurs after decision making. It is likely to be assumed that the decisions are right if they do not cause accidents, even when high-risk behavior occurs. This phenomenon is related to the "evaluation of emergency response" mentioned before. When nothing occurs in situations where people have made themselves sufficiently ready for a crisis, explanations such as, "It wasn't necessary to go to such lengths" are bound to have a certain influence.

In addition, group think also has the effect of rationalization in decision making. Even groups consisting of highly cohesive and talented members tend to fall into fixed viewpoints and thereby become unable to adopt flexible responses. Dangerous situations may not be correctly recognized when groups are mired in such pitfalls. Then a misperceived unanimity emerges and leads to further erroneous rationalizations.

The third devil's law includes the mistake of memory. Although our brains have enormous storage capacity, there is a limit. We cannot accurately recall all of our experiences. It is of no use to us to remember items that are not important or essential to our lives. We therefore tend to arrange information in our memory unconsciously. As a result, the phenomena of *incorrect remembering* and *forgetting* occur. In addition, the details of our memories might be affected by our desires and situation. We remember advantageous things and forget disadvantageous ones. Sometimes, the opposite can also happen. In either case, we do not even realize that our memories are incorrect. Also, the mechanism of forgetting comes into play when encountering negative experiences such as death, failures, etc. When sadness or regret is too mortifying to go on indefinitely, forgetting is one of the indispensable mechanisms for us to continue our lives well. At the same time, this mechanism can cause us to misjudge situations.

The Devil Creeping in on Our Carelessness: The Pitfall of Procedures

The probability of accidents is very low thanks to highly improved production systems and quality control. As a result, the devil creeps in slowly, giving us the mistaken belief that accidents almost never happen. This is the fourth devil's law. When serious accidents or disasters happen, incompleteness of the procedures can be the source of problems. However, in most cases, the rules and the procedures are made properly, especially involving situations with dangerous work. The problem occurs when workers do not follow procedure, even when there are visual signs and guidelines. The main reason for ignoring rules or procedures is the fact that accidents do not necessarily happen when the rules and procedures are not followed. In general, most machines and equipment are designed to ensure safety. Therefore, accidents will not happen unless we violate a rule or intentionally take unsafe actions. If the formula *fatal rule violation = accident* was generally correct, all rules and procedures would surely be followed. Nobody would exceed the speed limit on highways if it always resulted in an accident. In reality, we ignore rules and procedure in daily life when there is only a possibility, rather than a certainty, of an accident. And the range of deviations from the rules extends further by accumulation of evidence from situations where there were no consequences, until it becomes our customary daily behavior. Such deviate behaviors are strengthened by the repeated experiences that accidents or other problems do not occur. The fact of not having accidents in itself becomes the reward, and deviations are reinforced in the daily work process. Further rewards may include reductions in working time or workload. There is also the attraction of meeting the cost-reduction demands made by the upper management. In this way deviating behaviors will be reinforced by the receiving of constant rewards. Because such deviations also affect the other members of the group, it may lead to the formation of a group norm that disregards the procedures and rules. In order to overcome that, it becomes necessary to teach the reasons why the procedures and rules were made and to show examples of accidents that occurred due to a neglect of the procedures or rules.

In addition, there is a troublesome fifth devil's law. The truth is that accidents sometimes happen even when we observe the rules and follow the procedures. For instance, traffic accidents sometimes occur even when we are driving at legal speeds. With this kind of experience, we tend to think that it is no use to follow the rules. Thus, our tendency to disregard rules and procedures becomes stronger and stronger. We should think about the devil that is always trying to influence our behavior.

Hazards of Morality and Morale

When thinking about the safety of the organization it is not sufficient just to understand human nature and to be cautious so as not to be caught in dangerous pitfalls. Because interpersonal relationships influence our safety behavior in the organization, it is important to understand the effects of the group in order to prevent mistakes and accidents. We tend to associate safety with accidents or disasters immediately related to our own lives. However, it is actually necessary to promote a state where the members are working healthily and enthusiastically for the safety of the organization. It cannot be said that the organization is healthy if there are awkward interpersonal relationships, or if there is low motivation caused by increasing stressors. Such an organization is like a pressure cooker ready to explode, with an unexpected accident as a result. Indeed, when it reaches a critical point, the existence of the organization itself is endangered.

Moral hazards, or the collapse of morality, involve immoral and unethical behavior. The confidence of society is lost through such moral hazards, and well-known companies have disappeared due to this phenomenon. A well-established maker of dairy products in Japan caused a food poisoning accident and then did not take appropriate measures afterward. That year the company went more than $200 million into the red. It was the first time such a thing had happened since the establishment of the company. In another case, a car company failed to conduct a recall responsibly and went nearly $700 million into the red that year. In a third case, a food company that had mislabeled imported meat as domestic incurred losses of more than $200 million, and the company has since been driven out of business.

Such disgraceful outcomes are in some way the compensation for the injustice done, but these compensations may lead to the loss of jobs for employees, which is a further injustice (Sankei simbun, 2001). Corresponding to the concept of morality is also morale. Morale is a measure of motivation or satisfaction with work. If the morale of the members of an organization is high, then the organization is healthy and its targets can be smoothly achieved. Serious problems probably will not happen in such an organization. Unfortunately, we do not have empirical data that demonstrates the relationship between morality and morale. In addition, morality has various definitions. However, as shown in Figure 11.2, we have demonstrated the influence of interpersonal relationships and leadership on workers' morale in the workplace (Misumi, 1985). We have also found the same results with regard to the mental health of the subordinates, their satisfaction with their organization, with their colleagues, and with the communication in the workplaces. On the basis of these empirical data, we can infer that the morale of members can be heightened by improving the human side of the organization. And as a result, the organization can be saved from moral hazard. To achieve the safety of the organization, it is necessary to understand human behavior from a group aspect, interpersonal relationships and leadership, etc.

OBSTINACY AND THE SYSTEM THAT EVALUATES IT

To improve the safety of an organization, a kind of obstinacy or stubbornness related to the following of rules and procedures is strongly needed, as well as a system that can evaluate it. Organizations in which members mutually evaluate the degree to which rules are being followed can achieve desirable safety results. We need to create such an environment. Sometimes, of course, trouble or problems occur, even when the rules are being followed. In that case, it would be best to change or revise the procedures or rules. There are always procedures and rules with problems that are in need of improvement. To have such a system and such a climate plays an important role in maintaining safety in organizations. Otherwise, following rules becomes a truly meaningless behavior. Moreover, it is critical to be able to figure out mistakes and accidents caused when individuals violate the rules. Many problems are unintentionally caused by the actions and mistakes of individuals. However, it is also necessary to avoid the concealment of the mistake as much as possible. In that sense, it is indispensable for an organization to have a system and

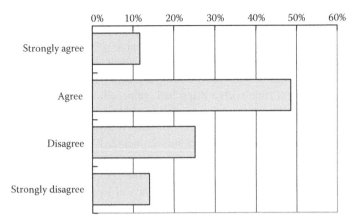

FIGURE 11.4   Perception of nurse leaders to the safety campaign and targets as mandatory.

an atmosphere where members can report their mistakes officially. Establishing a "no blame culture" is required.

As for the campaigns and targets, they will not work well unless they are accepted by all the members of the organization. Yoshida (2006) reported the results of a survey that consisted of twenty-five items about the targets and the safety campaign to ensure the safety for 333 nursing leaders. Included among the items were "there are a lot of rules and they are hardly followed," "the safety campaign turns into an external behavior and is seen as just a mere formality or ceremony," and "it is forced onto us as a matter of duty." More than 60% of the people answered, "strongly agree" or "agree" to the question, "Do you think the campaign and targets for safety are being forced on you?" (Figure 11.4).

No one will oppose attempts to try to decrease the number of accidents and mistakes, and objections are rarely made to the campaign to ensure safety. However, if the members of an organization do not accept the rules, the rules become mere "pie in the sky" and slogans, and will therefore end in failure.

AIMING TO ACHIEVE ORGANIZATIONAL SAFETY

We have emphasized, from the viewpoint of group dynamics, the importance of improvement of leadership and interpersonal relationships to improve the safety of an organization. We have also created scales to measure leadership and the safety environment in nuclear power plants. We then took the next step, the development and implementation of training to achieve an improvement in the leadership of those working as managers in nuclear plants (Misumi, 2001). These studies made clear that an improvement in leadership raises morale and has a positive effect on the safety awareness of subordinates. On the other hand, tendencies toward taking risky behaviors, which we refer to as the devil's laws, can emerge in groups and can reduce the power of the group itself. We also have implemented studies about so-called small-group activities as an effective approach to ensure safety (Koh, 1972). It was found that this would not work sufficiently when it was seen as an added new burden by members of the organization, and that it even could have the opposite effect. Members must recognize the small-group activities as a normal part of their work and as the best way to realize a safer working environment. The leaders in the workplaces can change the awareness of workers, and in this sense the leadership in an organization and the improvement of personal relationships play a big role. It is both an old and a new problem. It is fundamentally human beings that constitute an organization, and accidents occur due to multiple related factors. There is no single method that can absolutely guarantee safety. We therefore have to continue searching for an effective method. From such a viewpoint, we have done research and activities mainly focusing on leadership and development and practice of training. We also have been introducing small-group activities in order to revitalize the organization and raise the safety consciousness of working members. Yet it is not enough to promote the individual group activity to establish the safety of an organization. The concrete means to improve communication between groups in organizations must also be introduced. Here, we would like to propose the MP system as one of the methods that can realize it.

It is important for us to have eyes by which we can see the organization from the outside. But how can we get such eyes? We propose the introduction of the MP system (Yoshida, 2007), where MP stands for migratory person. These persons migrate through the organization like migratory birds. As the organization gets bigger, the individual sections get more independent, and members of each section tend to think their group is going well and do not care what is going on in the other groups. The groups become self-satisfied. This can result in a non-communicative situation between the groups, which may cause mistakes or accidents. Moreover, when a problem occurs, different groups tend to put the responsibility on one another. A wall that prevents mutual understanding is formed between them. Groups can also become self-centered. Facts may therefore be seen differently by the different groups. If the groups could recognize these points, communication between them would be improved rapidly. We propose introducing the MP system to overcome such problems. The core of this system is that members of the organization go through the workplace like migratory birds. As a result, chances to communicate increase and mutual understanding of different work groups improves. It is expected that the MP will communicate with members in all sections of an organization. Their role is not surveillance but rather to ask questions and to try to ensure safety. For example, they might ask about a machine at a site in terms of safety.

In the process of communication, "communalizing" and having a common understanding of information are indispensable. Likert (1961) emphasized the importance of the role of "linking pins" in the organization, and MPs serve basically the same role. In the case of linking pins, the role was limited to managers in the organization. In contrast, the MPs' work is carried out by members of the organization, regardless of their specific posts or previous experiences. They are appointed randomly and their role has a limited term. It is possible that a complete amateur will move throughout the organization. This could seem to lead to an increase in costs at first glance, but in the long term this system should promote communication

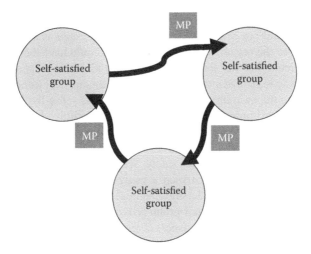

FIGURE 11.5   Introducing the MP (migratory person) system.

throughout the organization and bring positive results by ensuring the safety of the organization.

Such an MP system is shown in Figure 11.5. The point of this system is that the organization maintains its health by its own power. The standards for selecting MPs are very important and are as follows:

1. There should be at least two MPs.
2. They should include a young person and an experienced one. As a result, each one can see their target from different viewpoints. And it might be best to choose them from different workplaces.
3. The period of an MP's term should be relatively short, and they should be frequently replaced. Any position tends to gain too much power over a long term. If MPs end up becoming like spies or military police, the system will not work as hoped. From six months to a year might be a reasonable limit.
4. The MPs should be designated randomly. Thus, for example, a new employee doing clerical work potentially could be paired off with an experienced technician. They would go around the different workplaces and talk with all the members.

Through this process, the MPs become "the eyes from the outside" and may thereby be better able to make criticisms of the daily work and routines. However, their criticism should not be intended to be negative, but rather to positively support the members of the organization. Mutual understanding can be realized through this process. In this sense, the MP becomes a translator who bridges the gap between different workplaces. They would in a sense be not bilingual but multilingual.

When a safety problem occurs, pressure from the outside, especially the government, increases. As a result, regulations become strengthened. It is far more desirable to establish safety independently, by the organization itself, without having control taken away by the outside. The MP system could be one of the most effective methods for achieving this goal. At present, concrete implementations are still waiting to be made, but we hope to introduce this system in the future and promote real studies of its effectiveness.

REFERENCES

Koh, T. (1972). Practice and evaluation of safety campaign by participation. In J. Misumi (Ed.), *Leadership* (pp. 282–384). Tokyo: Diamond-Sha Publishing (in Japanese).

Lewin, K. (1947). Group decision and social change. In T. M. Newcomb & E. L. Hartley (Eds.), *Readings in social psychology* (pp. 330–344). New York: Henry Holt.

Likert, R. (1961). *New patterns of management.* New York: McGraw-Hill.

Misumi, J. (1985). *The behavioral science of leadership.* Ann Arbor, MI: University of Michigan Press.

Misumi, J. (Ed.) (2001). *The science of leadership and safety.* Kyoto: Nakanisiya Publishing (in Japanese).

Misumi, J. & Shinohara, H. (1967). The effects of group decision making on the prevention of accidents of bus drivers. *The Journal of Educational and Social Psychology, 6,* 125–134 (in Japanese).

Sankei simbun. (2001). *Why the excellent company failed?* Tokyo: Kadokawa Shoten (in Japanese).

Yoshida, M. (1999). Leadership training for organizational development. *Job Stress Research, 7,* 43–48 (in Japanese).

Yoshida, M. (2006). Safety in organization and understanding human nature. In T. Hasegawa (Ed.), *Encyclopaedia of safety in medical management* (pp. 22–26). Tokyo: Asakura-Shoten Publishing (in Japanese).

Yoshida, M. (2007). Safety management in organization and understanding human nature. *Nuclear Eye, 153,* 10–13. Tokyo: Nikkan-Kogyo-Simbun Publishing (in Japanese).

12 Organizational Influences on Safety Climate in Guanxi-Oriented Cultures

Shang H. Hsu, Chun-Chia Lee,
Muh-Cherng Wu, and Kenichi Takano

CONTENTS

INTRODUCTION

Researchers have recently recognized organizational factors as one of the important latent causes of failures in large, complex systems. Hollnagel & Woods (2005) have pointed out that 30–40% of incidents and accidents of large-scale complex systems relate to organizational factors. A series of empirical studies have found that organizational factors are linked to employee safety behavior (Neal, Griffin, & Hart, 2000; Seo, 2005; Tomas, Melia, & Oliver, 1999) as well as to organizational safety outcomes such as injuries, incidents, and accidents (Siu, Philips, & Leung, 2004; Varonen & Mattila, 2000). The ways in which organizational factors affect safety in high-risk industries have therefore become a critical issue in safety management (Flin et al., 2000).

Nowadays we are in the era of globalization. Manufacturing facilities around the world are integrated and tightly coupled. If an accident occurs, it may have a large-scale impact on the world's economy, politics, and society. Some industries, such as commercial aviation and nuclear power, attempt to apply successful safety improvement programs developed in one country directly to other countries. However, this migration process has not always been successful. For example, crew resource management (CRM) training courses, designed to improve communication and command styles in flight crews, have not been successful in Asian and South American countries. This lack of expected effect may be attributed to cultural differences. It is more precisely due to the fact that communication and command styles prescribed by CRM training are based on a Western-style culture, which assumes egalitarian and open communication. Applying these same assumptions to a culture that emphasizes hierarchy and order is very difficult. To be successful, a safety improvement program must accommodate the idiosyncrasies of its target country. It is therefore very important to understand the cross-cultural influence on safety management.

The workforces of the emerging Asian countries, especially the Greater China areas (including China, Hong Kong, and Taiwan) have recently attracted the attention of organizational researchers. One important aspect of national culture in Chinese Asia is the concept of *guanxi*, believed by many foreign managers and local Chinese leaders to be a critical success factor for business in the Chinese context (Tsui, Farh, & Xin, 2000). The term guanxi (關係), a particular kind of interpersonal relationship, refers to the implicit mutual interest and benefit a person develops in his or her social connections (Yang, 1994). This guanxi orientation is more important to most Greater China regions than in Western countries (Tsui & Farh, 1997; Xin & Pearce, 1996). Guanxi-oriented culture in Chinese context regions has also been described as a "Confucian dynamism" (Hofstede & Bond, 1988). The characteristics of a guanxi-oriented or Confucian dynamism culture greatly impact manager and employee decisions and behaviors in the Chinese context. Guanxi appears in both popular and academic literature because relationships based on guanxi are endemic in Chinese companies (Yang, 1994). Little attention, however, has been paid to how cultural differences in safety management might affect operational safety in high-risk industries, and particularly what the effects of the notion or concept of guanxi in Asia may be.

This chapter, therefore, focuses on the development of a guanxi-oriented safety management model in high-risk industries. The first section clarifies the concepts of guanxi orientation in the Chinese context. We then identify characteristics of guanxi-oriented cultures affecting organizational safety. Then, we propose a guanxi-oriented safety management model to describe the relationships between organizational factors of guanxi-oriented culture and safety behaviors and safety attitudes of workers in high-risk industries, and next present empirical test results for the model. Potential practical implications for prevention are discussed. Finally, the chapter offers suggested directions for future research.

GUANXI-ORIENTED CHARACTERISTICS IN THE CHINESE CONTEXT

Guanxi has many meanings in the Chinese language, referring more or less to the relationship between different perspectives. In this study, we focus on the work relationships within an organization. Work relationships in the Chinese context are built on family-like connections among people. The family-like connection is based on the demographic relationship (such as common background) as well as characteristics of the relationship itself (such as friendship) (Tsui & Farh, 1997). Yang (1994) indicated that guanxi refers to a dyadic relationship based implicitly on mutual interest and benefit. Members of guanxi-oriented cultures place very high priority on harmonious working relationships, loyalty, and maintaining hierarchical order (Hwang, 1987; Warner, 1995; Westwood, 1997). The great value placed on harmony exerts a strong influence on organizational structure, interaction among members, and the relationship between leaders and subordinates (Tsui et al., 2004; Westwood, 1997).

The guanxi-oriented culture exhibits several characteristics. First, guanxi-oriented organizations tend to be collectivist (Helmreich & Merritt, 1998; Hofstede, 1980) and emphasize hierarchical order (Hofstede, 1980, 1991; Westwood, 1997). A greater power distance exists between a superior and subordinates. Persons in such organizations place high value on organizational loyalty, obedience, social conformance, and compliance with authorities (Hofstede, 1980; Ho & Chiu, 1994). Guanxi-oriented organizations view obedience as both natural and proper (Westwood, 1997) and expect subordinates to respect authority and comply with superiors' directives (Farh & Cheng, 2000; Westwood, 1997).

Second, leadership style in a guanxi-oriented organization tends to be paternalistic. Paternalistic leadership style combines strong discipline and authority with fatherly benevolence and moral integrity (Farh & Cheng, 2000; Pye, 1981, 1985). A paternalistic leader is expected to exert centralized control and to be considerate, yet also to be directive and autocratic (Farh & Cheng, 2000). In guanxi-oriented societies, leadership is legitimized clearly by its members, who neither question nor challenge the rights of leaders (Westwood, 1997). In other words, leaders have absolute authority over subordinates and demand unquestionable obedience. Although paternalistic leaders tend to reprimand subordinates for poor performance (Farh & Cheng, 2000), they may not necessarily praise them for jobs well done. According to the tenets of Confucianism, however, an ideal paternalistic leader should demonstrate holistic concern for subordinates' personal and family well-being (Farh & Cheng, 2000).

Third, persons in guanxi-oriented organizations place high value on harmonious interpersonal relationships (Shi & Westwood, 2000; Ho & Chiu, 1994). Guanxi implies a special kind of relationship characterized by implicit rules of obligations and reciprocities (Hwang, 1987; Yeung & Tung, 1994), together with a necessary mechanism for coping with a highly non-codified social order and getting things done (Shi & Westwood, 2000). Maintaining in-group harmony is also highly valued in guanxi-oriented cultures (Westwood, 1997; Ho & Chiu, 1994). Harmony is achieved not through equality, but through accepting socially approved rules governing relationships between hierarchical levels. For example, one unspoken but accepted rule is that "face" (reputation, status, or dignity) of both superiors and subordinates should be respected and protected at all times (Westwood, 1997; Ho & Chiu, 1994). Another implicit rule that stipulates reciprocity and repaying favors also plays an important role in maintaining harmonious interpersonal relationships and governing exchanges (Hwang, 1987; Westwood, 1997; Ho & Chiu, 1994).

From a Western cultural perspective, the characteristics of guanxi-oriented culture seem to be negative to the safe operation of large, complex systems in the workplace. However, little research has discussed the relationships between the characteristics of guanxi-oriented culture and safety attitudes and behavior.

SAFETY MANAGEMENT IN GUANXI-ORIENTED CULTURE

Merritt & Helmreich (1996) conducted an extensive cross-cultural survey on the safety attitudes of airline pilots. They compared the cross-cultural differences of safety attitudes and behaviors between Asian and Western pilots and found most Asian pilots to be collectivists (i.e., less willing to stand out in a crowd and less likely to voice opposing opinions in meetings). A high power distance or hierarchical distance existed in Asian pilots' attitudes toward command, viewing superiors as the ones to take control in emergency situations and subordinates as the ones to follow without question. The researchers also found that Asian pilots avoid uncertainty by strictly adhering to organizational rules to make decisions and relying more on automatic devices to perform their duties. In contrast, Western pilots consider themselves more as individuals than as members of a group. Findings show a shorter power distance between captain and crew among Western airline workers and an open communication in both directions. Western pilots are also more likely to eschew organizational rules and exhibit less reliance on automation.

In a study of the safety climate among Chinese airlines, von Thaden and colleagues (2006) furthermore found that Chinese airlines score lower on accountability, employee empowerment, and active reporting compared to the findings of a previous study of Western airlines. Specifically, the Chinese airline pilots under study complained of favoritism and inconsistent standards and felt blamed for every accident and incident. Fearing blame, they were less likely to report their own mistakes as well as the mistakes of others. Pilots in these airlines did not feel empowered by management because they did not have the authority to make critical decisions.

Recently, Hsu et al. (2008) compared the safety climates of oil refineries in Taiwan and Japan and found that Taiwanese and Japanese safety climates differ in safety leadership, safety management approach, teamwork style, and safety behavior. Taiwanese upper management tends to be more top-down directive and is more actively involved in safety promotion than their upper management counterparts in Japan. Line managers in Taiwanese companies provide workers with more instructions and monitor work processes more closely. With regard to safety management, the Japanese managers tended to take a proactive approach involving preventive measures to reduce risks in the workplace, whereas the Taiwanese managers tended to be more reactive. That is, Taiwanese management usually takes remedial measures after an incident or accident occurs. Among the remedial measures, Taiwanese management often implements safety activities as a major means of promoting safety. Comparing Taiwanese and Japanese workers with regard to the overall climate, Taiwanese workers focus more on interpersonal relationships and place greater emphasis on a harmonious work atmosphere. They also abide by management directives without questioning managerial decisions.

The results of these studies suggest that the national culture has a strong influence on employees' safety attitudes and behavior. The national culture influence on workplace safety is becoming an increasingly important management issue with the growth in the number of overseas foreign-operated factories.

DEVELOPMENT OF A GUANXI-ORIENTED SAFETY MANAGEMENT MODEL

To effectively evaluate the impact of organizational factors on safety, one has to understand how an organization functions. An organization sets its goals and develops its strategies in response to requirements imposed by the changing environment. Top-level management makes policies to determine strategic goals and chooses the means to achieve those goals. Middle-level management formulates operating procedures to provide tactical policy action guidelines (Zohar, 2000). Line managers in the work group level execute policies and procedures, give directives to frontline workers, and supervise the work process to ensure safe and reliable operation (Zohar, 2000; Zohar & Luria, 2005). Zohar (2000) argued that group level factors mediate the influence of organizational level factors on individual safety behavior. Therefore, this study proposes a safety management model that adapts and extends the level concept assumptions of Zohar (2000) and postulates that organizational factors affect safety management and group-level process, and then in turn affect safety behavior and safety attitudes (see Figure 12.1).

This model divides the factors affecting organizational safety into four categories: organizational level factors, safety management level factors, work group level factors, and individual level factors.

- The organizational level includes three factors: upper management commitment to safety, tendency to blame others for mistakes, and emphasis on harmonious relationships. Management commitment to safety indicates the extent to which upper and middle-level management involve themselves in critical safety meetings and activities. Blame culture is the extent to which management blames employees for making mistakes or for unsafe behavior. Harmonious relationship refers to the extent of a harmonious work atmosphere among coworkers and supervisors in an organization.

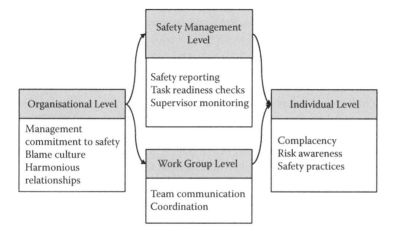

FIGURE 12.1   The conceptual model of this study.

- Safety management level factor categories include safety reporting, checking for task readiness, and supervisory monitoring. Safety reporting describes the willingness of employees to report safety issues honestly. Task readiness checks indicate the employees' perception of the line manager's check on task readiness before the task. Ongoing supervision refers to the extent to which supervisors continue to provide instruction and monitor employee safety.
- Work group level factor categories include team communication and coordination. Team communication is the extent of actively sharing and exchanging safety information among team members. Team coordination reflects the extent of collaboration among team members.
- Individual level factor categories include complacency, risk awareness, and safety practices. Complacency means the extent to which employees blindly follow management directives and procedures. Risk awareness indicates employee perception of risk at work. Safety practices refer to employee risk-taking and compliance with safety rules and procedures.

The detailed assumptions/hypotheses underlying this model can be described as follows. First, previous research suggests that management commitment is critical to employee safety performance (Zohar, 1980). Because upper management in Taiwan is expected to act like the father of a family, by setting an example for subordinates (Farh & Cheng, 2000) and by actively participating in safety activities and frequently promoting his or her own concern for safety (Hsu et al., 2008), greater upper management commitment to safety should influence the supervisory practices of the company's line managers, making them more serious about safety supervision, including task readiness checks and progress monitoring. An increase in task readiness checks or pre-task meetings, however, could lead to complacency. Workers may possibly think that if they pass the safety checks of their supervisors, then they will be safe. On the other hand, increased task supervisions might encourage employees to comply with safety procedures and regulations and therefore reduce worker complacency. The first part of the model consequently proposes that line managers in companies with higher management commitment to safety exhibit greater involvement in task readiness checks and ongoing supervision. More task readiness checks may result in greater employee complacency. However, more ongoing supervision may reduce employee complacency.

Furthermore, upper manager authority and power in a guanxi-oriented culture is accepted as natural, proper, absolute, paramount, unchallengeable, and inviolate (Farh & Cheng, 2000; Westwood, 1997). When upper managers are perceived as having a high commitment to safety, employees may be motivated to please upper management by only reporting good news and hiding the bad news. This type of selective reporting could lead to decreased employee safety awareness and deterioration in employee safety practices. Therefore, this model proposes more selective reporting in companies with higher management commitment to safety, which in turn reduces safety awareness and adherence to safety practices.

Second, researchers have suggested that a blame culture might discourage employees from reporting workplace safety problems and thus negatively impact employee safety performance (Reason, 1990, 1997). Employees in a patriarchal guanxi-oriented culture are often blamed or punished for their mistakes or for violating safety rules (Farh & Cheng, 2000). To avoid blame or punishment from management, employees may decide selectively what news to report, sharing the good news and hiding problems. As a consequence, upper management may not be aware of actual circumstances, which may lead to biased decisions regarding safety policies and procedures. Selective reporting may also keep other employees from becoming aware of the safety conditions of the workplace, hence making them less interested in implementing safety practices or following safety rules. Therefore, the second part of this model expects more selective reporting in companies that tend to blame or punish workers for their mistakes, leading to reduced safety awareness and reduced adherence to safety practices.

Third, previous studies have shown that maintaining a harmonious relationship is an important social value embedded in guanxi-oriented cultures (Westwood, 1997; Hwang, 1987; Ho & Chiu, 1994). Helmreich & Merritt (1998) found that Taiwanese pilots placed a high value on maintaining good relationships with managers and coworkers. Valuing harmonious working relationships encourages members of an organization to develop good interpersonal relationships and mutual trust (Tsui & Farh, 1997). Trust among employees should in turn facilitate group processes (e.g., people helping each other and openly sharing safety information) (Hsu et al., 2008). The likelihood of establishing a group level safety climate is greater with good information sharing and team collaboration (Lee & Harrison, 2000), in turn leading to increasing employee safety awareness and improving employee safety practices (Neal, Grant, & Hart, 2000; Seo, 2005; Tomas, Melia, & Oliver, 1999). Therefore, the third part of this model postulates that better communication and coordination among team members in companies with greater harmony among team members will enhance safety awareness and safety practices.

A harmonious work atmosphere, however, may have a negative impact on the reporting system. First, employees may report good news only in order to maintain harmony with superiors. Second, because reciprocity is highly valued (Hwang, 1987; Shi & Westwood, 2000), employees may cover up their coworker's mistakes, because reporting a colleague's mistakes to a superior would be considered a betrayal and unethical by coworkers. On the other hand, hiding colleagues' mistakes may also reduce employee awareness of safety problems and reduce compliance with safety practices. Therefore, this model expects employees to engage in more selective reporting of safety issues or mishaps in companies with greater harmony among team members, thereby reducing safety awareness and safety practices. Figure 12.2 provides a graphical depiction of specific relationships in this guanxi-oriented safety management model.

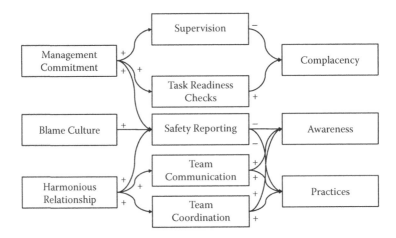

FIGURE 12.2   The guanxi-oriented safety management model.

EMPIRICAL RESULTS AND PRACTICAL IMPLICATIONS OF THE GUANXI-ORIENTED SAFETY MANAGEMENT MODEL

A safety climate survey questionnaire developed by the Central Research Institute of Electric Power Industry (CRIEPI) in Japan (Takano et al., 2004) was employed for this research. Six hundred and ninety front-line workers in high-risk industries (including chemical processing plants, steel foundries, and semiconductor foundries) in Taiwan, representing a guanxi-oriented culture (Hofstede & Bond, 1988), were used as samples.

Table 12.1 summarizes the descriptive statistics for each factor. Results of the descriptive analysis indicate that the workers rate higher on the factors of harmonious relationships and management commitment and lower on blame culture. The higher rating on management commitment indicates management gets deeply involved in safety-related activities and meetings. Under the circumstances, we found employees rated lower on reporting, which indicates employees selectively reported safety issues to management. The higher rating on harmonious relationships indicates that employees valued harmonious relationships among team members in the workplace. The scores for task

---

TABLE 12.1
Descriptive Statistics of Raw Scores among Factors

Factors	Mean	SD	Factors	Mean	SD
Management commitment	3.88	0.75	Team communication	3.77	0.63
Blame culture	3.52	0.81	Team coordination	4.00	0.54
Harmonious relationships	3.92	0.60	Reliance complacency	3.17	0.67
Task readiness checks	3.91	0.67	Risk awareness	4.23	0.48
Ongoing supervision	3.75	0.78	Safety practices	4.07	0.48
Safety reporting	3.51	0.61			

readiness checks were significantly higher than ongoing supervision. This indicated line management emphasized the importance of task readiness checks over supervision in the work process, which may have led to more reliance complacency of employees. The scores for risk awareness were higher than safety behavior, which implied employees' cognition of safety management was more than of safety behavior.

The study used structural equation modeling (SEM) to examine linkages among organizational factors, safety management factors, group factors, and workers' safety attitudes and behavior (Joreskog & Sorbom, 1993). Structural equation modeling is a statistical technique that allows assessment of both the direct and indirect effects of each variable on the other variables. There are a number of model fit indices, but no single fit index seems sufficient for a correct assessment of fit in an SEM model. In this study, we adopted several goodness-of-fit indices recommended by researchers to evaluate measurement adequacy (Joreskog & Sorbom, 1993; Bentler & Bonett, 1980): chi-square (χ^2), normed fit index (NFI), non-normed fit index (NNFI), comparative fit index (CFI), incremental fit index (IFI), and root-mean-squared error of approximation (RMSEA). Bentler (1992) recommends an NFI, NNFI, CFI, and IFI of .90 or greater as an acceptable data fit. An RMSEA value up to .05 indicates a good model fit; a value of .08 or less indicates a reasonable model fit; a value greater than .10 indicates poor model fit (Joreskog & Sorbom, 1993). The overall fit indices of the structural model were assessed by χ^2 (684) = 3381.37, $p < .01$. The values of RMSEA were .076 (less than .08), indicating the structural model was a reasonable fit. Other indices (NFI = .90; NNFI = .91; CFI = .92; IFI = .92) were greater than or near .90, indicating that the structural model was acceptable. In summary, test results indicate that the structural model is adequate.

Results of the SEM examination are summarized as follows:

- In companies with higher management commitment to safety, line managers will be more involved in task readiness checks. In companies with more task readiness checks, there will be greater employee complacency.
- In companies with higher management commitment to safety, line managers will be more involved with ongoing supervision. In companies with more ongoing supervision, there will be less employee complacency.
- In companies with higher management commitment to safety, there will be more selective reporting. In companies with more selective reporting, safety awareness and the adherence to safety practices will be reduced.
- In companies with greater harmony among team members, there will be better communication among team members. In such companies, safety awareness and safety practices are enhanced.
- In companies with greater harmony among team members, there will be better coordination among team members. In such companies, safety awareness and safety practices are enhanced.
- In companies with greater harmony among team members, employees will engage more in selective reporting of safety issues or mishaps. In companies in which there is more selective reporting, safety awareness and safety practices are reduced.
- Finally, contrary to expectations, the tendency to blame others for mistakes does not significantly effect how safety issues are reported.

These results are in line with those reported by other organizational behavior studies on guanxi-oriented companies. Traditional social values, social harmony, and reciprocity in guanxi-oriented culture play important roles in influencing safety management and employees' safety attitudes and behavior (Hsu et al., 2008; von Thaden et al., 2006). Workplace harmony shapes group processes. Harmonious interpersonal relationships enhance trust among team members and their leaders (Tsui & Farh, 1997). Team members in harmonious atmospheres are more willing to share information and communicate with each other. Furthermore, the existence of harmonious interpersonal relationships in guanxi-oriented cultures obligates team members to coordinate with each other and reciprocate favors (Hwang, 1987; Tsui et al., 2004). Better communication and coordination improve the safety climate, in turn enhancing risk awareness and safety behavior (Hofmann & Stetzer, 1996). Workplace harmony yields high-quality leader–member exchange, which can facilitate good communication with higher management (Hofmann & Morgeson, 1999).

The current study developed several implications for improving safety in paternalistic guanxi-oriented companies. First, safety training should incorporate relationship building. Building a good relationship is a prerequisite for improving teamwork effectiveness in relationship-oriented organizations. Good interpersonal relationships can enhance both trust and group cohesiveness, which in turn facilitate group communication and coordination. To build a good relationship in the workplace, leaders should actively promote corporate identity of employees in addition to maintaining group harmony. At the same time, management should assist employees to honor team performance and place higher priority on team performance than their own. By doing so, it will be easy to foster interpersonal trust among team members. Trust in management and coworkers will not only promote the quality of teamwork, but also enhance safety performance of employees in the workplace.

Second, harmonious relationships also have a detrimental effect on safety management; i.e., with greater harmony among team members, employees will tend to be more selective in reporting on safety issues, which in turn reduces safety awareness and safety practices. The trade-off between the advantages and disadvantages of a harmonious relationship is a critical issue. To mitigate the detrimental effect, leaders should clearly define job tasks to avoid the unfair treatment of employees; for example, workers who have a good relationship with the leader may be given privileges or responsibility for important functions. Management should make clear rules and regulations for the work environment and build a good accountability system in which the organization rewards safe behavior and dispenses consequences for unsafe behavior. Reason (1990) indicated that one of the most important safety climate factors is the manner in which both safe and unsafe behavior is evaluated and the consistency of assessing rewards and penalties according to their evaluations. Therefore, management may build a good reward system based on team performance to avoid employees covering mistakes for each other. To prevent human-made manipulation for performance, management may provide unbiased performance standards and evaluation tools, such as automatic computer-based checklists. In addition, building a good report culture is an important issue. An effective and systematic reporting system is key to identifying the weaknesses and vulnerabilities of safety management before an accident occurs. Reporting systems might be better focusing on workplace hazards rather than human errors. Employees in harmonious organizations may avoid mentioning other persons involved in an incident, as a means of saving face.

Third, although task readiness checks are necessary, they should be accompanied by increased ongoing supervision. To avoid the reliance complacency of employees, line managers should be visible in the workplace reviewing safety-related activities, enforcing safety job procedures, and training employees in safe job procedures. They may need to follow work processes more closely, provide instructions, and routinely remind workers of safety rules. Management should build a mechanism, through implementing self-check and double-check protocols, to encourage employees to improve safety performance. Line managers should also participate in staff training courses at which safety policies and procedures are introduced and reviewed. In addition, line managers should reinforce the continuous improvement attitudes of employees. Lack of continuous improvement attitudes will reduce the capabilities of organizational change and encourage reliance complacency, which is so harmful to the development of a safety climate.

CONCLUSION

The guanxi-oriented culture is prevalent in Greater China areas. We find little attention in prior research focused on how cultural differences in safety management in guanxi-oriented cultures might affect operational safety in high-risk industries. Although Merritt & Helmreich (1996) have conducted a cross-cultural survey comparing the differences of safety attitudes and behavior between Asian and Western pilots, it is limited to a single aviation-based study. We do not know whether it can be applied to other domains. Therefore, this work has attempted to examine the generalizability of their study to other application domains. The findings of this work are consistent with those of Merritt & Helmreich's study. That is, Taiwanese process industries have a strong tendency of guanxi-oriented characteristics. Specifically, we find the workers' efforts to keep harmonious relationships are a very important feature in Taiwanese workplaces, one of important social value in guanxi-oriented culture. Upper management tends to be actively involved in safety promotion activities in order to demonstrate their high commitment to safety. Furthermore, front-line workers involved in an incident or accident are less likely to be blamed for their errors because of paternalistic leadership.

This study further explores the relationships among organizational factors, safety management, team process, and the individual's safety behavior in the context of guanxi-oriented culture. It finds that leadership style and organizational atmosphere significantly influence safety management and work group process, in turn influencing individual safety attitudes and behavior. That is, paternalistic leadership, different from Western leadership, attempts to set a safety model and demonstrate high commitment to safety by high involvement in safety activities. A leader's high commitment to safety will influence the individual worker's safety awareness and behavior. In addition, we find that a harmonious relationship influences the extent of communication and coordination in workplaces, which in turn will influence the individual worker's safety awareness and behavior. Furthermore, we also find that harmonious relationships can have both an incremental and a detrimental effect on safety management, depending on the line manager's supervisory activity.

The findings of this study can be used as a basis for improving safety management programs in guanxi-oriented cultures. Suggestions for safety management are proposed, based on observations of the safety culture reform programs of several outstanding companies in Taiwan (e.g., Taiwan Semiconductor Manufacturing Company, EVA Air, etc.). Most important, a fair and just culture should be established in the workplace (Reason, 1997). Management should not give special favor to people who are close to them. Tasks should be assigned to competent workers. Safe and unsafe behavior should be clearly defined. Ongoing supervision and quality checks should be enforced during the work process. Task performance should be measured in an unbiased manner, and unambiguous feedback should be given to the work team. Rewards should be contingent on team performance, rather than being personal favors. By doing so, safety management can leverage positive effect while minimizing the negative effect of the organizational factors inherent in guanxi-oriented culture.

This study has presented a first look at safety management in guanxi-oriented cultures. Future researchers must extend its generalizability to other regions of guanxi-oriented cultures. In addition, more organizational characteristics should be included in the proposed model in order to enhance its validity.

REFERENCES

Bentler, P. M. (1992). On the fit of models to covariances and methodology to the Bulletin. *Psychological Bulletin, 112,* 400–404.

Bentler, P. M. & Bonett, D. G. (1980). Significance tests and goodness-of-fit in the analysis of covariance structures. *Psychological Bulletin, 88,* 588–606.

Farh, J. L. & Cheng, B. S. (2000). A cultural analysis of paternalistic leadership in Chinese organizations. In J. T. Li, A. S. Tsui & E. Weldon (Eds.), *Management and organizations in the Chinese context* (pp. 84–127). New York: Macmillan.

Flin, R., Mearns, K., O'Connor, P. & Bryden, R. (2000). Measuring safety climate: Identifying the common features. *Safety Science, 34,* 177–193.

Helmreich, R. L. & Merritt, A. C. (1998). *Culture at work in aviation and medicine: National, organizational, and professional influences.* Aldershot, UK: Ashgate.

Ho, D. Y. F. & Chiu, C. Y. (1994). Component ideas of individualism, collectivism, and social organization: An application in the study of Chinese culture. In U. Kim, H. C. Triandis, C. Kagitcibasi, S. C. Choi & G. Yoon (Eds.), *Individualism and collectivism: Theoretical and methodological issues* (pp. 137–156). Thousand Oaks, CA: Sage.

Hofmann, D. A. & Morgeson, F. P. (1999). Safety-related behaviour as social exchange: The role of perceived organizational support and leader–member exchange. *Journal of Applied Psychology, 84,* 286–296.

Hofmann, D. A. & Stetzer, A. (1996). A cross-level investigation of factors influencing unsafe behaviours and accidents. *Personnel Psychology, 49,* 307–339.

Hofstede, G. (1980). *Culture's consequences: International differences in work-related values.* Newbury Park, CA: Sage.

Hofstede, G. (1991). *Cultures and organizations: Software of the mind.* New York: McGraw-Hill.

Hofstede, G. & Bond, M. H. (1988). The Confucius connection: From cultural roots to economic growth. *Organizational Dynamics, 16,* 4–21.

Hollnagel, E. & Woods, D. D. (2005). *Joint cognitive systems: Foundations of cognitive systems engineering.* Boca Raton, FL: CRC Press.

Hsu, S. H., Lee, C. C., Wu, M. C. & Takano, K. (2008). A cross-cultural study of organizational factors on safety: Japanese vs. Taiwanese oil refinery plants. *Accident Analysis and Prevention, 40,* 24–34.

Hwang, K. K. (1987). Face and favor: The Chinese power game. *American Journal of Sociology, 92,* 944–974.

Joreskog, K. & Sorbom, D. (1993). *LISREL 8: Structural equation modeling with the SIMPLIS command language.* Chicago: Scientific Software International.

Lee, T. & Harrison, K. (2000). Assessing safety culture in nuclear power stations. *Safety Science, 34,* 61– 97.

Merritt, A. C. & Helmreich, R. L. (1996). Human factors on the flight deck: The influences of national culture. *Journal of Cross-Cultural Psychology, 27,* 5–24.

Neal, A., Griffin, M. A. & Hart, P. M. (2000). The impact of organizational climate on safety climate and individual behaviour. *Safety Science, 34,* 99–109.

Pye, L. W. (1981). *Dynamics of Chinese politics.* Cambridge, MA: Gunn and Hain.

Pye, L. W. (1985). *Asia power and politics.* Cambridge, MA: Harvard University Press.

Reason, J. (1990). *Human error.* New York: Cambridge University Press.

Reason, J. (1997). *Managing the risks of organizational accidents.* Aldershot, UK: Ashgate.

Seo, D. C. (2005). An explicative model of unsafe work behaviour. *Safety Science, 43,* 187–211.

Shi, X. & Westwood, R. I. (2000). International business negotiation in the Chinese context. In J. T. Li, A. S. Tsui & E. Weldon (Eds.), *Management and organizations in the Chinese context* (pp. 185–221). New York: Macmillan.

Siu, O. L., Philips, D. R. & Leung, T. W. (2004). Safety climate and safety performance among construction workers in Hong Kong: The role of psychological strains as mediators. *Accident Analysis and Prevention, 36,* 359–366.

Takano, K., Tsuge, T., Hasegawa, N. & Hirose, A. (2004). Development of a safety assessment system for promoting a safe organizational climate and culture. In N. Itoigawa, B. Wilpert & B. Fahlbruch (Eds.), *Emerging demands for nuclear safety of nuclear power operations: Challenge and response* (pp. 57–71). Boca Raton, FL: CRC Press.

Tomas, J. M., Melia, J. L. & Oliver, A. (1999). A cross-validation of a structural equation model of accidents: Organizational and psychological variables as predictors of work safety. *Work and Stress, 13,* 49–58.

Tsui, A. S. & Farh, J. L. (1997). Where guanxi matters: Relational demography and guanxi in the Chinese context. *Work and Occupations, 24,* 56–79.

Tsui, A. S., Farh, J. L. & Xin, K. R. (2000). Guanxi in the Chinese context. In J. T. Li, A. S. Tsui & E. Weldon (Eds.), *Management and organizations in the Chinese context* (pp. 225–244). New York: Macmillan.

Tsui, A. S., Wang, H., Xin, K. R., Zhang, L. & Fu, P. P. (2004). Let a thousand flowers bloom: Variation of leadership styles among Chinese CEOs. *Organizational Dynamics, 33,* 5–20.

Varonen, U. & Mattila, M. (2000). The safety climate and its relationship to safety practices, safety of the work environment and occupational accidents in eight wood-processing companies. *Accident Analysis and Prevention, 32,* 761–769.

von Thaden, T. L., Li, Y. J., Li, J. & Lei, D. (2006). *Validating the commercial aviation safety survey in the Chinese context.* Technical Report HFD-06-09, Human Factors Division. Savoy, IL: University of Illinois.

Warner, M. (1995). *The management of human resources in Chinese industry.* London: Macmillan.

Westwood, R. (1997). Harmony and patriarchy: The cultural basis for "paternalistic headship" among the overseas Chinese. *Organization Studies, 18,* 445–480.

Xin, K. R. & Pearce, J. L. (1996). Guanxi: Connections as substitutes for formal institutional support. *Academy of Management Journal, 39,* 1641–1658.

Yang, M. M. (1994). *Gifts, favors and banquets: The art of social relationships in China.* Ithaca, NY: Cornell University Press.

Yeung, Y. M. & Tung, R. L. (1994). Achieving business success in Confucian societies: The importance of guanxi (connections). *Organizational Dynamics, 25,* 54–65.

Zohar, D. (1980). Safety climate in industrial organizations: Theoretical and applied implications. *Journal of Applied Psychology, 65,* 96–102.

Zohar, D. (2000). A group-level model of safety climate: Testing the effect of group climate on microaccidents in manufacturing jobs. *Journal of Applied Psychology, 85,* 587–596.

Zohar, D. & Luria, G. (2005). A multilevel model of safety climate: Cross-level relationships between organization and group-level climates. *Journal of Applied Psychology, 90,* 616–628.

13 A Comprehensive Evaluation of Licensees' Efforts on Safety Culture

Yoichi Ishii and Maomi Makino

CONTENTS

INTRODUCTION

In the beginning of nuclear power plant development, technological problems and individual problems were the main sources of operational failures and disturbances. With the upgrade of the plants and the consequent increase in complexity, the human–machine interface became a problem. Later, after the Chernobyl accident, the safety culture became a big problem. In recent years the occurrence of organization-related accidents has continued in Japan, such as the JCO criticality accident (NSC, 1999), the TEPCO false records problem (NISA, 2002), and the Mihama accident (NISA, 2004).

In this chapter we describe the recent trend in efforts to improve the safety culture in many countries and in the International Atomic Energy Agency (IAEA). In 1991, the IAEA for the first time reported on the nuclear safety culture in the world, and since then various reports and suggestions have been provided. The IAEA has developed the safety culture evaluation items and provided the evaluation technique and the viewpoints for evaluation to all member countries. Our report makes reference to these aspects. The recent Safety Standard GS-R-3 considers safety culture as the framework of the integrated management system. The IAEA has also begun a new safety review service, called SCART, an abbreviation for Safety Culture Assessment Review Team. As a safety review service, SCART reflects the expressed interest of member countries for methods and tools for safety culture assessment. The IAEA safety fundamentals, requirements, and safety standards are the basis of the SCART safety review service. SCART missions are based on the SCART guidelines (SCART, 2008), which provide overall guidance to ensure the consistency and comprehensiveness of the safety culture review. SCART guidelines were established in February 2008.

In the United States, the safety culture is not regulated directly but is inspected indirectly in the ROP (reactor oversight process). For example, the self-evaluation of safety culture is required if repeated human errors of the same type are caused by the cross-cutting issues in ROP.

In the countries of the European Union, the regulatory requirements for the "management system for safety" are being provided according to the IAEA framework to regulate the safety culture. In France, the safety culture assessment has started to be included in the regular inspection program.

RECENT SAFETY CULTURE TRENDS IN JAPAN

The development in the number of reports of legally notified events per operating unit and the ratio of human error events to the total number of reports are described in Chapter 8. The figure there shows that whereas there is a decrease in the number of incidents, the rate of human errors is increasing rather than decreasing. This is because something called organization-related events is included in these human errors. The recent organization-related incidents in Japan are listed in Table 13.1.

The main organization-related problems pointed out in the reports from these incidents are as follows:

- Exclusivity of organizations
- Insufficient provision of information

TABLE 13.1

Recent Japanese Organization-Related Accidents

Year	Event	Description
1995	Accident of sodium leakage from the PNC's* FBR** Monju	The sodium leakage occurred because of damage to the thermometer. The damage was due to failure of the thermometer sheath and inappropriate information handling. (NSC, 1995)
1997	Fire and explosion accident in PNC's asphalt solidification facility	The fire was extinguished with water spray when it occurred. However, the explosion on the following day was due to inadequate extinction of the fire. (NSC, 1997)
1998	Falsification of data on spent fuel transport containers by Nuclear Fuel Transport Ltd. and Nuclear Works Ltd.	It was found that there had been intentional data falsification concerning a spent fuel transport container's neutron shielding materials. (STA, 1998)
1999	Critical accident in JCO uranium processing plant	During uranium refining work, the uranium nitrate solution in a precipitation tank reached a critical state, resulting in radiation exposure of 59 JCO personnel, including the death of two JCO employees and at least seven residents in the vicinity. (NSC, 1999)
2002	Discovered problem of false self-checking record by Tokyo Electric Power Co.	It was discovered that self-controlled-inspection records had been falsified by the Tokyo Electric Power Co. The records concerned the core shroud and the containment leak inspection. (NISA, 2002)
2004	Rupture of piping in the secondary system in Mihama Unit 3, Kansai Electric Power Co.	There was a pipe rupture accident of the secondary system at Unit 3 of the Mihama Power Station of the Kansai Electric Power Co. The rupture was caused by reduced pipe strength due to wall thinning by so-called erosion/corrosion at the downstream side of an orifice in the secondary system piping. (NISA, 2004)
2007	Revealed critical accident in Shika Unit 1, Hokuriku Electric Power Co.	In 2007, a criticality accident and unexpected dislodgement of control rods at Unit 1 of Shika Nuclear Power Station, Hokuriku Electric Power Co. was noted to have occurred on June 18, 1999, during a period of maintenance outage. The cover-up of the accident and data falsifications were not disclosed for 8 years. (NISA, 2007a)

* Power Reactor and Nuclear Fuel Development Corporation.
** Fast breeder reactor

- Insufficient experiences and knowledge, and problems in the succession of technologies and skills
- Basic recognition fades with time
- Problems in the compliance of laws and ordinances and the maintenance of records
- Insufficient understanding of (the importance of) information disclosure.

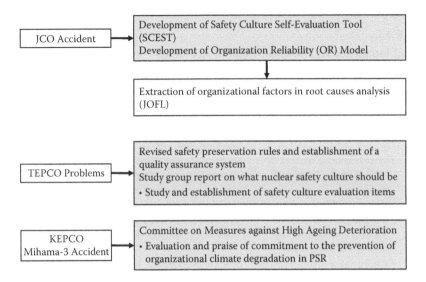

FIGURE 13.1   JNES commitments to organizational accidents and events.

- Paralyzed quality assurance function in the whole company
- Defects in safety culture

Although the nuclear industry in Japan in many ways has reached maturity, new problems caused by the degradation of safety culture still occur. Many of these are related to organizational problems.

CURRENT APPROACH FOR ORGANIZATION-RELATED PROBLEMS IN REGULATIONS

Figure 13.1 shows the JNES commitments to organizational accidents and events in Japan. After the JCO accident, the TEPCO problem, and the Mihama accident, the JNES investigated and developed the existing tools for evaluating safety culture, such as the safety culture self-evaluation tool, or SCEST, the OR model, and the JOFL (JNES, 2006), among others. Details about this can be found in Makino, Sakaue, & Inoue (2005).

Applied to the Mihama-3 accident, the JNES commitments lead to the following considerations. As has already been described in Chapter 1, the direct cause of the Mihama-3 accident was the deterioration of equipment due to aging. But this took place against the background of a defective safety culture. As part of improved measures against aging deterioration, the regulatory body therefore in 2005 proposed the following steps to prevent the degradation of the corporate culture and organizational climate:

- The licensee should confirm their efforts for building of safety culture as a part of their quality assurance activities in their periodical safety reviews. The periodical safety reviews are carried out every ten years.

- The regulatory body should assess the efforts made by the licensees and should support them by actively praising all good cases.
- The efforts should be assessed according to the following ten key issues (JNES, 2005):

1. Commitment by top management
2. Clear policies being executed by supervisors
3. Improvement and establishment of a QMS (quality management system)
4. Reporting culture and a habitual questioning attitude
5. Learning organization
6. Working place with good communication
7. Eliminating decision making due to misjudgment
8. Observing rules
9. Accountability and transparency
10. Self-assessment or assessment by a third party

NEW APPROACH FOR ORGANIZATION-RELATED PROBLEMS IN REGULATIONS

It has been pointed out in a report from NISA (2006) that human errors often have been due to organizational factors or the degradation of safety culture and organizational climate. To further improve the safety level of nuclear power facilities, it is therefore necessary to thoroughly correct the nonconformity that is associated with direct factors such as human errors, organizational factors, the degradation of safety culture, and organizational climate, and to establish three guidelines as follows:

1. Guidelines for assessing the evaluations and improvements made by the licensee that directly affect human factors and human errors
2. Guidelines for root cause analysis
3. Guidelines for how to prevent the degradation of safety culture and organizational climate

In addition, the previously mentioned NISA report (2006) made clear that "to ensure its more reliable commitment to enhance safety, it is important for any licensee to make efforts that prevent the safety culture and climate of its organization from being degraded" (pp. 11–12).

In addition to including such efforts in the conventional "periodical safety review" (every ten years), it is also necessary to evaluate the licensee's efforts to preserve safety in daily activities.

To bring such an approach into practice, the JNES began the preparation of the guideline for the prevention of safety culture and organizational climate degradation. The JNES further established a study committee of safety culture in the safety regulation to discuss the contents of the guideline. The members of this study committee included experienced scientists, university professors, and licensees, as well as representatives from NISA and JNES. The committee held eight meetings to work on the guideline.

During the drafting process for the guideline, fourteen items were proposed as the components of safety culture, hence as the viewpoints from which to evaluate the efforts made to prevent a degradation of safety culture and organizational climate in the day-to-day safety preservation activities. As already described in a previous section, the JNES defined ten key issues to capture the efforts to cultivate a safety culture in the periodical safety reviews. These ten items were derived from a survey of the following twelve main literature sources and were considered to be suitable in Japanese culture:

1. *Management of Operational Safety in Nuclear Power Plants* (INSAG, 1999)
2. *Key Practical Issues in Strengthening Safety Culture* (INSAG, 2002)
3. *Guidelines for Organizational Self-Assessment of Safety Culture and for Reviews by the Assessment of Safety Culture in Organizations Team* (IAEA, 1994)
4. ASCOT guidelines, revised 1996 edition (IAEA, 1996)
5. *Developing Safety Culture in Nuclear Activities: Practical Suggestions to Assist Progress* (IAEA, 1998)
6. *Self-Assessment of Safety Culture in Nuclear Installations: Highlights and Good Practices* (IAEA, 2002a)
7. *Safety Culture in Nuclear Installations: Guidance for Use in the Enhancement of Safety Culture* (IAEA, 2002b)
8. *Safety Culture in the Maintenance of Nuclear Power Plants* (IAEA, 2005)
9. *IAEA Report of the OSART Mission to the Kashiwazaki-kariwa Nuclear Power Plant* (OSART, 2004)
10. *The Role of the Nuclear Regulator in Promoting and Evaluating Safety Culture* (NEA, 1999)
11. *Report of Committee for Implementation of Nuclear Safety Culture in Japan* (NUPEC, 2003)
12. *Safety Culture Study Group Report* (SCSG, 2000)

"VIEWPOINTS FOR EVALUATION" ACCORDING TO THE FOURTEEN COMPONENTS OF SAFETY CULTURE

The JNES reviewed the ten items according to new findings from the Davis-Besse accident report (U.S. Nuclear Regulatory Commission, 2004), the Space Shuttle accident report (CAIB, 2003), and others, and then defined the fourteen components of safety culture for the daily preservation activities. The detailed contents of the fourteen items are described in the following sections.

The NISA and the JNES prepared this guideline, the *Guideline for the Regulatory Body to Evaluate the Licensee's Efforts to Prevent Degradation of Safety Culture and Organizational Climate* (NISA, 2007b), to be used by inspectors for the new safety preservation inspection. The NISA further ordered licensees to add the provision of developing a safety culture to their safety preservation rules. The inspection for the new safety preservation rules started in December 2007, and the guideline went into effect at the same time.

COMMITMENT BY TOP MANAGEMENT

The first and vital component in achieving excellent safety performance is a commitment to safety and to the strengthening of a safety culture at the top of an organization. This means that safety clearly and unequivocally is put in first place in the requirements from the top of the organization. It also means that there should be absolute clarity about the organization's safety philosophy. We therefore defined the following two items as important aspects of this component:

- The top management's clear message to give top priority to safety is well disseminated even to the lowest level of the organization.
- Top management should actually do the action according to their message to give top priority to safety. By this message and action, all members of the organization can work without creating conflict between safety targets and the targets to pursuing financial or business interest, etc. Therefore, this message should include such content.

CLEAR POLICIES AND EXECUTION BY SENIOR MANAGERS

The senior management of the organization should establish a value system that makes the significance of the safety preservation activities clear. It is important to make the organization members behave so that they give priority to the safety goals, without feeling any conflict between the goal for safety and those for different objectives (such as pursuing profit or maintaining electric power supply). We therefore defined the following three items as important aspects of this component:

- Presentation and implementation of a policy for activities to ensure safety
- Establishment and execution of resource planning (including the budget plan, staffing plan, equipment investment plan, and maintenance plan), with top priority given to safety (including the correction for deviation of the plans from the priorities corresponding to safety, significance, urgency, etc.)
- Establishment and full functioning of systems in charge of the safety preservation activities of the entire organization (head office and power station), as well as the role, responsibility, and authority of each department

MEASURES TO AVOID WRONG DECISION MAKING

In situations in which group think is widespread in an organization, deviating views may be suppressed. Even if individual members of a group have objections or doubts about a certain assertion, they may tend to think that their doubts will be found inappropriate and that it will be more beneficial to go along with the opinions of the group. This does not mean that a group member cannot express a doubt that he or she feels, but rather that the communication of it is paralyzed by group think. It is clearly necessary to find ways of preventing a collective thoughtlessness from taking place. We therefore defined the following two items as important aspects of this component:

- Specific measures should be established to avoid erroneous decision making on safety and to eliminate isolationism in the organization (collective thoughtlessness, etc.), and it should be ensured that they are functioning well.
- With regard to safety preservation activities, the decision-making system defined by the quality management system should be followed. For instance, Chapter 4.1 of ISO.9001 (General Requirements) stipulate that the organization shall: a) identify the processes needed for the quality management system and their application throughout the organization, b) determine the sequence and interaction of these processes, and c) determine criteria and methods needed to ensure that both the operation and control of these processes are effective. In accordance with this, operating process sequences and maintenance process sequences are determined by the organization and documented in, e.g., operating manuals and maintenance manuals.

HABITUAL QUESTIONING ATTITUDE

A well-tested system that relies on defense-in-depth and is supported by procedural requirements will protect employees and the public from radiation hazards. Because of this, it is easy for the workforce to develop the attitude that safe conditions are provided for them by someone else. They may also think that events at other plants are exceptional and isolated occurrences that could not occur at their own plant. In order to prevent that, it is essential that everyone connected with nuclear safety is constantly reminded of the potential consequences of failing to give safety absolute priority. Most incidents and accidents in the nuclear industry have occurred because someone has failed to take the relevant precautions or has failed to consider or question, in a conservative manner, decisions that were made or the steps that were taken to implement them. We therefore defined the following item as an important aspect of this component:

- All members of the organization should establish a habitual questioning attitude toward their own behavior, the status of equipment, and the way the organization works, as this pertains to safety.

REPORTING CULTURE

An organization with a good safety culture regards failures and near-misses as good lessons that are useful to avoid more serious events. There is thus a strong motivation to ensure that all events that are potentially instructive are reported and investigated to discover their root causes. The motivation also means that timely feedback is given on the findings and remedial actions, both to the work groups involved and to others in the organization or the industry who might experience the same problem. We therefore defined the following item as an important aspect of this component:

- It is important to cultivate an atmosphere in the workplace such that individual errors, near-misses, and other information that may seem to be undesirable for the organization can be reported without undue concerns. Moreover, the senior manager must take the lead in this by providing a role model.

GOOD COMMUNICATION

It is important to establish a tight communication system for the enhancement of safety culture. Any gaps in understanding between the administrative management staff and young employees, as well as between the administrative management staff and technically responsible employees, must be filled. We therefore defined the following two items as important aspects of this component:

- The in-house communication must function effectively, between organizational levels (superiors and inferiors) as well as across the organization.
- The dialogue with contractors and the notification of requirements to contractors should be conducted appropriately, and the notifications should be well known. Efforts should also be made to create communication opportunities in order to promote a mutual understanding.

ACCOUNTABILITY AND TRANSPARENCY

Accountability and transparency are important to obtain understanding and acceptance of local residents, the public, and the regulatory body. When providing information, it is essential to set up a system to analyze and understand the information needs of local residents and the public. We therefore defined the following item as an important aspect of this component:

- When a situation that requires an explanation arises, transparent information should be provided to local residents, the public, and the regulatory body in a timely manner. Moreover, efforts should be made to create communication opportunities to promote a mutual understanding.

COMPLIANCE

If rules (such as procedures) are not properly valued, it is inevitable that shortcuts or workarounds will become a part of practice. This could lead to further degradation of the safety level, because working around a requirement, which is not itself a prime safety requirement, quickly can lead to a culture in which even vital and fundamental safety rules are no longer viewed as sacrosanct. In order to enhance safety culture it is therefore necessary that an indictment system is established in case compliance issues occur. We therefore defined the following three items as important aspects of this component:

- To manage and maintain rules (procedures) in order to ensure that they are appropriate and effective. This should include timely reviews, revisions or annulment, new developments, etc.

- To establish compliance in routine work.
  The purpose of compliance is to observe legislation, regulations, and regulatory requirements in order to realize the objectives of the organization and to comply with the internal company rules (e.g., standards, requirements, and procedures related to nuclear safety) so as to achieve nuclear safety in response to the social demands underlying legislation and regulations.
- To establish and cultivate a system and an atmosphere that make it possible to deal with problems effectively with respect to compliance.

LEARNING ORGANIZATION

In a learning organization, it is possible to pick up ideas, energy, and concerns from people at all levels of the organization. Enhancements in safety are sustained by ensuring that the benefits obtained from improvements are widely recognized by individuals and teams, and this in turn leads to even greater commitment and identification with the process of improving safety culture. We therefore defined the following four items as important aspects of this component:

- Maintain and improve the technical capabilities of the organization by fostering and motivating members at each level of the organization, including top management and managers, through training and education, competence evaluation, selection/qualification, etc.
- Accumulate and disseminate the knowledge, information, and data related to safety preservation activities to the departments concerned.
- Acquire and accumulate knowledge from important in-company and domestic as well as overseas accidents and failures. The study of and reflection on this knowledge should lead to corrective actions.
- Acquire and accumulate knowledge from human errors and near-miss analyses. The study of and reflection on this knowledge should lead to corrective actions.

ORGANIZATION COPING WITH PREVENTION OF ACCIDENTS AND TROUBLES

In order to prevent accidents and problems, and to maintain safety, the organization should have methods for root cause analyses as well as corrective action programs. We therefore defined the following item as an important aspect of this component:

- In order to prevent accidents, failures, and the like, the knowledge acquired from the root cause analyses of the accidents and failures should be disseminated to the organization. Examples of this are nonconformity management, corrective actions, preventive measures, etc.

SELF-ASSESSMENT OR THIRD PARTY'S ASSESSMENT

To prevent safety culture promoting activities from becoming ineffective, the organization should have available a self-assessment or third-party assessment. These assessments should be kept independent and prevented from becoming routine exercises. We therefore defined the following two items as important aspects of this component:

- To prevent the safety culture promoting activity from becoming ineffective, a self-assessment or third-party assessment should be performed.
- Indicators for gauging the achievement level of safety culture cultivating and for detecting symptoms of degradation of safety culture should be defined, and the self-assessment should be performed.

WORK MANAGEMENT

To improve the work environment, it is important that the organization have a reasonable schedule for planning and implementing appropriate fieldwork. We therefore defined the following item as an important aspect of this component:

- The organization should have a reasonable schedule for fieldwork and implement fieldwork so that it does not lead to overwork. The organization should improve environments and implement fieldwork under clean and safe conditions.

CHANGE MANAGEMENT

It is important that the organization perform adequate control of changes in organization and rules. We therefore defined the following two items as important aspects of this component:

- For any changes in the organization (including changes related to contractors), an appropriate assessment of their risk and safety impact as well as adequate change control should be carried out.
- For any changes in rules, procedures, etc., an appropriate assessment of their safety impact as well as adequate change control should be carried out.

ATTITUDE AND MOTIVATION

It is necessary for the organization to make efforts to improve employees' attitude and motivation in order to prevent the degradation of safety culture. We therefore defined the following three items as important aspects of this component:

- The organization should make efforts to improve employees' willingness and attitude, to heighten their motivation and to make their workload reasonable for their routine work.
- The organization should make efforts to improve managers' leadership, willingness, and attitude for management.
- The organization should make efforts to foster a good workplace climate.

REFERENCE CRITERIA FOR THE FOURTEEN COMPONENTS OF SAFETY CULTURE

It is not easy to develop clear reference criteria for these fourteen components, so we have developed examples of a degraded safety culture for each of the components. Using these, it should be relatively easy to perform the evaluation for a specific component. Some examples of how this can be done are provided in Table 13.2.

TABLE 13.2
Examples of Bad Safety Culture Cases

Component	Examples of Inappropriate Compliance or Bad Safety Culture
Good communication	The in-house exchange of safety information is not regular. Moreover, neither the information transmission route nor the means in the organization are regularly reviewed.
	The majority of the communication is in one direction, from the top management, and communication from the employee level is insufficient.
	Neither the cooperative relationship nor the communications from the regulator or the representative of licensees is sufficient.
Commitment by top management	There is lack of clear organizational commitment to safety.
	There is lack of management awareness and involvement in plant activities.
	The organization is subjected to increasing economic and market pressures. The organization's prime policy is to meet short-term profit goals.
Organization coping with prevention of accidents and troubles	Design and equipment modifications often stall because attention is given to "firefighting" rather than to addressing root causes.
	There is large backlog of inoperable equipment.
	There is a significant accumulation of corrective actions that have not been addressed.
Learning organization	There is inadequate radiological training of workers.
	The effect of the education and training is not regularly evaluated.
	There is no attitude of learning from other organizations to improve the business process. Even when workers learn from other organizations, they are not praised.
Habitual Questioning attitude	When problems occur, there is a highly ritualized response to them.
	There is self-satisfaction with current performance; no need to look for problems.
	There is no STAR (stop, think, act, review) attitude.

It is naturally necessary that these fourteen components are continuously reviewed as part of their actual use and that this review incorporates the latest knowledge of the discipline.

METHODS TO COMPREHENSIVELY EVALUATE THE EFFORTS TO PREVENT DEGRADATION OF SAFETY CULTURE AND ORGANIZATIONAL CLIMATE

As already mentioned, the NISA started the new inspection for safety culture. The inspectors evaluate the licensee's safety culture according to a method consisting of the following ten steps, using the viewpoints of the fourteen components of safety culture:

1. Survey of the action plan and indicators of the licensees' safety culture cultivating activities

   In the beginning of each fiscal year, the inspector surveys the action plan and indicators of the licensees' safety culture promoting activities and enters the survey results on the evaluation sheet shown in Table 13.3.

2. Survey of the licensee's selection of symptom indicators of the degradation of safety culture

   At the beginning of each fiscal year, the inspector surveys the indicators selected by the licensee as useful for evaluation of symptoms of degradation of safety culture, and enters the survey results on the evaluation sheet shown in Table 13.3, with survey frequency.

3. Extraction and arrangement of matters considered to be symptoms of degradation of safety culture and organizational climate from observation of the licensee's daily preservation activities

   The inspector takes hold of the actual situation of operation, maintenance, etc., of a nuclear power plant through routine patrols, participation in the licensee's meetings for operational safety activities, periodic inspections, pre-service inspections, safety preservation inspections, etc., and enters the apparent symptoms of degradation of safety culture and organizational climate on the evaluation sheet shown in Table 13.3. This must include reasons for their judgment of points of view that evaluate degradation indications of safety culture components.

4. Extraction of matters related to root cause analysis results concerning the safety culture

   When an event occurs in the nuclear power plant and the root cause is analyzed, the inspector enters the matters considered to be symptoms of the degradation of safety culture and organizational climate from the results of licensee's root cause analysis on the evaluation sheet shown in Table 13.3. This must include their reasons for the various symptoms.

5. Survey of the results of the licensee's efforts for safety culture

   At the end of each fiscal year, the inspector surveys the results of licensee's efforts for safety culture according to the action plan and indicators

Table 13.3

Comprehensive Evaluation Sheet on Efforts to Prevent Degradation of Safety Culture and Organizational Climate

Name of electric company / Power station	(Step 10)	
Period for evaluation: From Month / Day / Year to Month / Day / Year From / / to / /	Comprehensive Findings	

Licensee's efforts for cultivating safety culture activities	Observation and findings (Step 3)	
	Safety culture component	Observation and findings, and their reasons
1. Action Plan (Step 1- (1) and (2))		
2. Indicators to evaluate activity results (Step 1- (3))		
3. Results of efforts (Step 5)		

Indicators for indirect evaluation of symptoms of degradation of safety culture		Results of root cause analysis (Step 4)	
Item & frequency	Survey results	Event	Matters and reason
(Step 2)	(Step 6)		

Selection of items considered necessary to enhance licensee's efforts for safety culture cultivating activities (Step 7)

Item requesting licensee's efforts, based on discussion with the licensee (Step 8)

Good Practices (Step 9)

surveyed in Step 1 and enters the survey results on the evaluation sheet shown in Table 13.3.

6. Survey of measured results of the indicators of the symptoms

At the end of each fiscal year, the inspector studies the licensee's survey results of the indicators in order to check the trend of indicators that are entered at Step 2 and considered useful for evaluation of symptoms of degradation of safety culture. The inspector enters the results in the evaluation sheet shown in Table 13.3.

7. Selection of items considered necessary to enhance the licensee's efforts

Based on the results of Steps 1 to 6, the inspector selects the items considered necessary to enhance the licensee's efforts from a perspective of preventing the degradation of the safety culture and the organizational climate. The inspector enters the selected items on the evaluation sheet shown in Table 13.3.

8. Presentation of items requesting the licensee's efforts

After sufficient discussions with the licensee, the inspector presents to the licensee the items, taken from the set developed in Step 7, considered to require enhancement of the licensee's efforts. The inspector enters the selected items on the evaluation sheet shown in Table 13.3.

9. Selection of good practices

The inspector selects the licensee's efforts suitable for encouragement as good practices and enters them on the evaluation sheet shown in Table 13.3.

10. Comprehensive findings for the safety culture

Based on the contents surveyed in Steps 1 to 9, the inspector produces his or her view of the licensee's comprehensive findings regarding efforts to prevent degradation of safety culture and organizational climate and the evaluation of symptoms of degradation of safety culture and organizational climate. These comprehensive findings should be described according to the following.

With regard to efforts to prevent degradation of safety culture and organizational climate, the inspector describes the situation according to the following four grades:

1. No efforts are made.
   Examples:
   − No involvement of top management.
   − No concrete action plan established.

2. Efforts are made but no improvements are observed.
   Examples:
   − Inadequate involvement of top management.
   − Efforts are limited to particular departments and individuals.
   − No improvement in the observed value of evaluation indicators.

3. Efforts are made based on the plan, and trends of improvements are observed.
   Examples:
   - Involvement of top management.
   - Efforts are made as those of the entire power plant according to the action plan.
   - Trends of improvements in the observed values of evaluation indicators.

4. Continuous improvements have been made.
   Examples:
   - Active involvement of top management.
   - Knowledge on factors or problems common to the entire power plant is accumulated.
   - Efforts and review of evaluation indicators are voluntarily made for continuous improvements.

Regarding the evaluation of symptoms of degradation of safety culture and organizational climate, the inspector describes the situation according to the following four grades:

1. Clear symptom of degradation of two or more safety culture components
2. Symptom of degradation of particular a safety culture component
3. Continuous monitoring required in order to observe further trends
4. Improvement trends observed; however, continuous monitoring is desirable without complacency

CONCLUSION

The inspector surveys and evaluates the licensee's efforts for cultivating the safety culture according to the ten steps described in NISA's *Guideline for the Regulatory Body to Evaluate the Licensee's Efforts to Prevent Degradation of Safety Culture and Organizational Climate*. It is very important that the inspector have active discussions with licensees and that he or she accepts the diversity in licensee's efforts and takes these views into consideration when making a comprehensive evaluation of licensee's efforts. The guideline is based on the fourteen components of safety culture described in this chapter, so this chapter mainly describes the viewpoints needed to make a comprehensive evaluation of the licensee's efforts for cultivating a safety culture. The authors expect this guideline to contribute to the further improvement of the safety culture of the Japanese nuclear industry and to reduce accidents caused by organizational factors.

ACKNOWLEDGMENT

The authors would like to express their special thanks to the staff of the Nuclear Power Inspection Division of NISA and to the members of the study committee of safety culture in the safety regulation for supplying valuable information and suggestions, and encouraging us to make this guideline.

REFERENCES

CAIB (Columbia Accident Investigation Board) (2003). *Report* (6 vols). Washington, DC: Government Printing Office.

IAEA (International Atomic Energy Agency) (1994). *Guidelines for organizational self-assessment of safety culture and for reviews by the assessment of safety culture in organizations team* (TECDOC-743). Vienna: IAEA.

IAEA (International Atomic Energy Agency) (1996). *Guidelines for organized self-assessment of safety culture and for reviews by the assessment of safety culture in organizations team. Revised 1996 edition* (TECDOC-860). Vienna: IAEA.

IAEA (International Atomic Energy Agency) (1998). *Developing safety culture in nuclear activities: Practical suggestions to assist progress* (IAEA Safety Report Series no. 11). Vienna: IAEA.

IAEA (International Atomic Energy Agency) (2002a). *Self-assessment of safety culture in nuclear installations: Highlights and good practices* (TECDOC-1321). Vienna: IAEA.

IAEA (International Atomic Energy Agency) (2002b). *Safety culture in nuclear installations: Guidance for use in the enhancement of safety culture* (TECDOC-1329). Vienna: IAEA.

IAEA (International Atomic Energy Agency) (2005). *Safety culture in the maintenance of nuclear power plants* (IAEA Safety Report Series no. 49). Vienna: IAEA.

INSAG (International Nuclear Safety Advisory Group) (1999). *Management of operational safety in nuclear power plants* (INSAG-13). Vienna: International Atomic Energy Agency.

INSAG (International Nuclear Safety Advisory Group) (2002). *Key practical issues in strengthening safety culture* (INSAG-15). Vienna: International Atomic Energy Agency.

JNES (Japan Nuclear Energy Safety Organization) (2005). *Concept of challenge and view point of understanding for prevention of deterioration of organizational culture.* JNES-SS report JNES-SS-0514. Tokyo: JNES.

JNES (Japan Nuclear Energy Safety Organization) (2006). *The safety culture assessment method (The implementation manual).* JNES-SS report JNES-SS-0616. Tokyo: JNES.

Makino, M., Sakaue, T. & Inoue, S. (2005). Toward a safety culture evaluation tool. In N. Itoigawa, B. Wilpert & B. Fahlbruch (Eds.), *Emerging demands for the safety of nuclear power operations* (pp. 73–84). London: CRC Press.

NEA (Nuclear Energy Agency) (1999). *The role of the nuclear regulator in promoting and evaluating safety culture.* Paris: Organization for Economic Cooperation and Development/NEA.

NISA (Nuclear and Industrial Safety Agency) (2002). *The interim report on the issues on falsification of self-controlled-inspection work records, etc.* Tokyo: NISA.

NISA (Nuclear and Industrial Safety Agency) (2004). *Final report on the pipe rupture accident of Unit 3 of the Mihama Power Station, the Kansai Electric Power Co., Inc. issued in March of 2005.* Tokyo: N ISA.

NISA (Nuclear and Industrial Safety Agency) (2006). *Improvement of the inspection system for nuclear power facilities* (pp. 11–12). Tokyo: Nuclear and Industrial Safety Agency.

NISA (Nuclear and Industrial Safety Agency) (2007a). *Investigation report on the criticality accident and unexpected dislodgement of control rods at Unit 1 of Shika Nuclear Power Station, Hokuriku Electric Power Co. in 1999.* Tokyo: NISA.

NISA (Nuclear and Industrial Safety Agency) (2007b). *Guideline for the regulatory body to evaluate the licensee's efforts to prevent degradation of safety culture and organizational climate* (NISA-166c-07-10). Tokyo: NISA.

NSC (Nuclear Safety Commission of Japan) (1995). *Nuclear Safety Commission report on February 19, 1995.* Tokyo: NSC.

NSC (Nuclear Safety Commission of Japan) (1997). *Nuclear Safety Commission report on December 22, 1997.* Tokyo: NSC.

NSC (Nuclear Safety Commission of Japan) (1999). *Report by the investigation committee for the criticality accident issued in December of 1999.* Tokyo: NSC.

NUPEC (Nuclear Power Engineering Corporation) (2003). *Report of committee for implementation of nuclear safety culture in Japan.* Tokyo: NUPEC.

OSART (Operational Safety Assessment Review Team) (2004). *IAEA report of the OSART mission to the Kashiwazaki-kariwa nuclear power plant.* Vienna: IAEA.

SCART (Safety Culture Assessment Review Team) (2008). *SCART (safety culture assessment review team) guidelines.* Vienna: IAEA.

SCSG (Safety Culture Study Group) (2000). *Safety culture study group report.* Tokyo: High Pressure Gas Study Institute of Japan.

STA (Science and Technology Agency) (1998). *Report by the investigation and study commission for spent-fuel transport container on December 3, 1998.* Tokyo: Science and Technology Agency.

U.S. Nuclear Regulatory Commission (2004). *Davis-Besse nuclear power station: NRC special inspection. Management and human performance corrective action effectiveness report* (no. 05000346/2004013). Washington, DC: U.S. Nuclear Regulatory Commission.

14 Hindsight and Foresight

Erik Hollnagel and Hiroshi Sakuda

CONTENTS

HINDSIGHT

One thing that characterizes industrial environments is that they become ever more complex. This complexity arises from a number of developments, many of which depend upon and affect each other. As far as the technology is concerned, it becomes more powerful, cheaper, and more reliable. The simplest illustration of that is Moore's law, the basic thrust of which is that everything — in the technological world — changes exponentially. As far as the industrial processes and services are concerned, there is an increasing horizontal and vertical integration — partly because of the possibilities that new technologies offer. This is seen not only within specific industries, but also in the way that, for example, energy distribution or finance links regions and countries. With regard to the demands from end users, be they owners or consumers, the requirements for faster, more diversified, and more reliable services seem insatiable. This means that products and services almost literally must be available anytime and anywhere. In recent years it has also become clear that all this must be achieved without undue effects on the local and global environments. These developments, and there may be others, contribute to the complexity of industrial environments.

The developments of the industrial environments take place continuously and lead to increased demands for their safe and efficient use. The improvements in our ability to manage and control these environments, however, are much smaller and are discrete rather than continuous. This is clearly seen in the developments of human factors but can also be recognized in other ways; for instance, the ability to manage the environment, public services, or the economy. In the case of human factors, the period from the beginning of the 1990s to today, roughly the lifetime of the INSS, has seen a number of changes, many of which have been described by the chapters in this book. The two most important ones are recounted in the following.

THE CHANGING VIEW OF HUMAN ERROR

Even though the accident at Three Mile Island in 1979 made the importance of the human factor clear, at least in the nuclear industry, other industries did not initially share the same concern. Nuclear power production was always acknowledged as having the potential for a disaster, and the concern for safety was ever present. Accidents could obviously also happen in other industries, but because such accidents usually only had local consequences, the risks were not as prevalent. The general perception of the human factor as a source of risk, in the form of human errors, gradually became part of the common wisdom to the extent that human error was accepted as a satisfactory explanation for 80–90% of industrial accidents. This perception is still commonly found and has spread to other domains such as traffic safety and patient safety.

Although the preference for a single — and simple — cause is a pervasive phenomenon in human thinking, it was gradually realized that it was inadequate to single out the human factor, in the form of human error, as a specific cause. Authors such as Woods & Cook (2002) have eloquently and convincingly argued for the need to look for "second stories," i.e., to go beyond the initial explanations or causes. Even though there still is widespread interest in finding effective ways to prevent human error, it is also accepted that this can rarely be done simply by preventing or eliminating specific human actions or operations. For instance, Hirotsu (in Chapter 6), points out that:

> (H)uman performance is not a simple matter. Problems in human actions are often influenced by combinations of organizational factors such as work planning and training, and work environment factors such as the on-site situation. To develop effective countermeasures, it is therefore essential to clarify problems and causal factors of events from the aspects of human factors. However, relationships among the causal factors have become more and more complicated because organizations and facilities get increasingly complex, and it has become increasingly difficult to identify fundamental causes.

The importance of the working situation and the social environment is recognized by many of this book's authors. It is perhaps expressed most clearly by Le Bot, who in Chapter 4 notes that:

> From our experience, individual human error is not predominant, nor does it explain the failures. In each failure scenario, the system is solid and redundant enough for the consequences of an individual error to be contained. Such an isolated error cannot lead to failure given the barriers and redundancies in place. For example, an operator's procedural error may be corrected by another operator or the supervisor who observes anomalies in their procedures. If the whole team is involved in an erroneous operation on the basis of an inadequate procedure, this operation cannot be taken to be a simple error of procedure implementation; it can arise only from a conscious operation of the team, which has decided on this course of action for good reasons due to the particular characteristics of the situation. This action may have been originally initiated by an individual error of one of the team members, but this cause is not sufficient. So we consider this error, which contributes to the generation of the final operation, as a feature of a special situation (or context). This situation or context should be analyzed in as much detail as possible so as to explain that the whole system may be involved in this erroneous operation.

Indeed, the presentation of the safe regulation model is an argument for considering the social environment in particular, rather the organizational environment in general. Humans base their work on a common understanding, not only of what happens in the industrial environment (the processes they deal with), but also of what is needed for the social group or collectivity (see Chapter 10), to achieve its objectives. Work is both constructed in a social environment and also depends on others meeting their obligations. Human performance can therefore provide a decidedly positive contribution to the safety of work, as described by Yoshida in Chapter 11:

> We would like to emphasize the importance of workers having the idea and sensitivity towards the concept of feeling unsafe. Technology should first of all be "fail-safe," but humans should have the "feel-unsafe" attitude at the same time. The concept of feel-unsafe is extremely important for sensing risk. It goes without saying that people should make choices based on the situation that lead to a more secure outcome, or are fail-safe. The concept of feel-unsafe will be useful in these situations, because technology, such as machinery, cannot feel unsafe. In contrast, humans have the power to feel unsafe by utilizing knowledge and experience gained over time. This can, of course, also include intuition. It is in principle conceivable that machines will have this sort of ability at some point in the future. However, even then, the ability of humans to feel unsafe will continue to play an important role for the realization of safety. It is thus desirable to adopt fail-safe for design and operation of equipment or devices and feel-unsafe for the people who use such equipment or devices.

CHANGING VIEWS ON SAFETY CULTURE AND ORGANIZATIONAL CULTURE

Changing the view of the human factor, and of human error, to recognize the importance of working conditions and the social environment quite naturally leads to a concern for what is usually referred to as organizational culture, and specifically safety culture. Organizational culture became a central topic in the field of organizational studies and management in the early 1980s and is often associated with the work of Edgar Schein. The concern for how the members of an organization characterize their environment, and how this environment affects their behavior, is of course as old as the study of administrative behavior itself (e.g., Simon, 1945/1997). Coincidentally, safety culture became an important topic at about the same time, partly as a means to understand the disaster at the Chernobyl nuclear power plant in April 1986.

The safety culture is important both in the way it may affect the actions of operators at the sharp end and in the way it may affect the performance of the organization as a whole. The former is obvious, both in the changing perspective on human error and in the interest for how different management styles can affect performance, and thereby also safety. In Chapter 5, Shibaike and Fukui explain it thus:

> These accidents were not just the consequences of hardware problems but also resulted from complex interrelated social and organizational factors. The present conditions can no longer be fully addressed by the human–machine interface approach alone. Under these circumstances, the workers' sense of values, their ways of thinking, their behavioral patterns and habits may hold the key to preventing accidents. This is what

we call the safety culture of the organization to which people belong. Safety culture is created within the context of relationships among people. The activities described in this chapter represent efforts to create a highly trustful working environment within the context of relationships among members of an organization.

However, as the accidents at the Mihama nuclear power plant show, safety culture is also important for the organization as a whole. This means not only that so-called blunt end (see the discussion in Hollnagel, 2004), but also the general way in which people — managers, administrators, and supervisors as well as operators — behave in everyday situations. Several of the chapters in this book refer to a number of cases in Japan, where organizational inadequacies or the lack of a safety culture have played a role. Similar cases can easily be found in other countries around the globe. More important, several chapters also describe ways in which the safety culture can be improved. In Chapter 5, Shibaike and Fukui argue that:

> … ensuring that individuals consider safety and safe behavior is one of the ultimate goals in preventing human errors, and it is an organizational environment that guides people to that goal. Thus, a safety climate can be defined as an organizational environment that guides its members to consider safety and safe behavior. Whereas safety culture encompasses the traits of an organization and of individuals, safety climate focuses not on individuals but on an organizational environment. Safety culture is built on a safety climate. This perspective of a safety climate is needed to cultivate a safety culture.

Safety culture has also become an important issue for the regulators, in nuclear power as well as in aviation and patient safety, to name a few.

The common approaches to safety culture are heavily influenced by the Western way of thinking. It is, however, possible to take a different approach, as described by Hsu et al. in Chapter 12. They introduce the Chinese notion of *guanxi*, defined as a kind of interpersonal relationship that refers to the implicit mutual interest and benefit a person develops in his or her social connections. The guanxi-oriented organization has three main characteristics that sets it apart from an occidental tradition: (1) guanxi-oriented organizations tend to be collectivist, (2) leadership style in a guanxi-oriented organization tends to be paternalistic, and (3) persons in guanxi-oriented organizations place high value on harmonious interpersonal relationships achieved by accepting socially approved rules governing relationships between hierarchical levels. Field studies have found that a leader's high commitment to safety does influence the individual worker's safety awareness and behavior. A harmonious relationship also influences the extent of communication and coordination in workplaces, which in turn will influence the individual worker's safety awareness and behavior.

FORESIGHT

This book, among other works dealing with the same issues, has clearly demonstrated that human factors have changed over the last few decades. The interesting question, for research and practice alike, is of course what the challenges are that lie ahead, and what changes they require. That there will be challenges is beyond

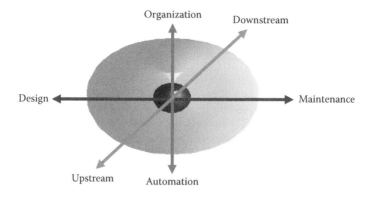

FIGURE 14.4   The extension of human factors.

a doubt. This is a consequence of the continued development of technologies and organizations. One way of illustrating that is proposed by Figure 14.1.

In Figure 14.1, the small sphere illustrates the focus of human factors around 1984, and therefore also the scope of work at the time. The focus was on a single type of work, typically operations. In some cases, the focus could be on, for example, design or maintenance, but it would be limited to a single part of work. The focus was on work where it took place; that is, at a single point of time in the production process, whatever it was. And the focus was on the work at the sharp end, although influenced by what was coming from the blunt end (i.e., something that took place at a different time and in a different place).

The larger, gray sphere illustrates how the focus of human factors has changed. This change can be described in terms of three extensions. One (horizontal) extension includes larger parts of the system's lifecycle, from design to maintenance. This recognizes the fact that whatever happens at the actual operations can be affected by what happened when the system was designed and what may happen during maintenance activities. A second, horizontal extension covers both upstream and downstream processes. This recognizes the fact that the level of integration is much larger, and that it is insufficient to consider the local work situation. Events from upstream may constrain current operations, for instance, in terms of delays or unexpected resource problems. And events downstream may set requirements that have to be accommodated in order to prevent future problems. The final extension is the vertical one, covering the entire system from the technology (automation) to the organization and management.

It seems inevitable that these developments will continue to take place and that the scope of human factors therefore must grow. Or rather, that if we want to be able to ensure the safety and efficiency of complex industrial environments in the future, then we must develop models and methods that consider the systems such as they will be. This can mean either that human factors changes its scope or that new hybrid fields of study — and fields of practice — emerge. Several of the chapters in this book illustrate how that can happen.

One specific development needed is that the understanding of accidents changes from one of root causes to more complex explanations. In Chapter 7, Furuhama points out that:

> The strict RCA [root cause analysis] idea presumes that true root causes exist, as causes without which the events would not have occurred. This assumption is incompatible with many actual accidents for which various causality relations among background factors have been found to explain the final consequences. We therefore think that the strict RCA idea is not adequate for practical use, and we accordingly use the term "root cause" in more general interpretation. Our emphasis in practice is on "cause" rather than on "root," which means that the whole set of background factors in itself should be understood as the root cause. Based on this idea, we will continue … training activities to promote our systematic ability to perform RCA.

Although for practical reasons it may be useful to keep using the term "root cause" (even though, strictly speaking, it is undesirable from a scientific point of view), it is necessary that the scope of causes is extended considerably to include organizational causes such as inappropriate management, production pressures, inadequate safety culture, etc. Because such causes are of a different type than equipment malfunctions or even human errors, they also require different responses. Such causes represent relations rather than functions and can therefore not be acted upon directly. It may even be difficult to suggest responses that directly address the supposed causes; how, for instance, can a safety culture be changed?

The extension of human factors to consider more complex phenomena also means that it is necessary to revise the traditional distinction between correct and incorrect, as, for example, in success or failure. If a process or phenomenon is complex, it quite basically means that it can be in many different states, and not just in two. Although an outcome may possibly be classified as either good or bad — provided that clear acceptability criteria can be agreed upon — the underlying process cannot.

The logical consequence of this is to accept that a complex process can give rise to several different types of outcomes, but that the process itself is the same. This means that the same complex process can either go right or go wrong. This viewpoint is expressed by resilience engineering (see Chapter 3), which looks at failures as the flip side of successes. Adopting this view also means that it is impossible to avoid accidents by dealing with the root causes. Because the same processes are behind successes and failures, constraining or eliminating these processes will also adversely affect the ability to succeed (i.e., to be safe and efficient).

Another way in which the scope of human factors must grow is to consider the interplay with society (Chapter 2). Safety and risk is no longer an issue just of the operations in front of the process. Nor is it only an issue of the organization, its internal processes, and its culture. The organization is itself part of an environment, which is the society. This society has expectations about how organizations and people work, how they prioritize various goals, and how much effort should be put into achieving them. The effect on the operator at the sharp end certainly may be indirect, but there is an effect nonetheless. When human factors start to consider such issues, it is perhaps no longer warranted to call it human factors. Be that as it may, the pursuit of safety for complex industrial environments should not be limited by semantics. The human factors of the future, say of 2020, may be a quite different discipline than the human factors of, say, 1980 or 2000. But so will the processes and working environments that it must deal with. It would not only be naïve but also dangerous and irresponsible to rely on the models and methods of the past. It is, indeed, the everlasting dilemma of human factors that it creates the complexity of tomorrow by solving the problems of today with the mindset, models, and methods of yesterday.

REFERENCES

Hollnagel, E. (2004). *Barriers and accident prevention*. Aldershot: Ashgate Publishing Limited.
Simon, H. A. (1945/1997). *Administrative behavior*. New York: The Free Press.
Woods, D. D. & Cook, R. I. (2002). Nine steps to move forward from error. *Cognition, Technology, and Work, 4*(2), 137–144.

Index